Control
of Insect Behavior
by
Natural Products

Control of Insect Behavior by Natural Products

Edited by
David L. Wood
DEPARTMENT OF ENTOMOLOGY AND PARASITOLOGY
UNIVERSITY OF CALIFORNIA
BERKELEY, CALIFORNIA

Robert M. Silverstein
DEPARTMENT OF CHEMISTRY
STATE UNIVERSITY COLLEGE OF FORESTRY
AT SYRACUSE UNIVERSITY
SYRACUSE, NEW YORK

Minoru Nakajima
DEPARTMENT OF AGRICULTURE CHEMISTRY
KYOTO UNIVERSITY
KYOTO, JAPAN

ACADEMIC PRESS
New York · London
1970

ACADEMIC PRESS, INC.
111 Fifth Avenue, New York, New York 10003

United Kingdom Edition published by
ACADEMIC PRESS, INC. (LONDON) LTD.
Berkeley Square House, London W1X 6BA

LIBRARY OF CONGRESS CATALOG CARD NUMBER: 69-13486

PRINTED IN THE UNITED STATES OF AMERICA

298666

CONTENTS

v

CONTENTS

CONTENTS

PREFACE

The extensive use of pesticides by the more affluent nations of the world has not only created serious problems in pest control but threatens man's health and pollutes his environment. Insecticides in particular are of greatest concern because they can cause death through poisoning, accumulate in man, concentrate in food chains, are often not biodegradable, cause resurgence and resistance in pest populations, and destroy parasites, predators, and pollinators. In the past few years, research aimed at establishing alternative means of pest control has received increased attention. One of the most promising of these is the use of naturally occurring organic compounds that influence insect chemosensory behavior as attractants, repellents, stimulants, deterrents, and arrestants. Many reports have established the presence of such chemosensory behavioral systems, and of particular interest is the number of serious economic pests included in this list.

However, significant advances in pest control utilizing this biochemical approach remain painstakingly slow because of our primitive understanding of insect behavior, problems associated with mass rearing and with isolation and identification of compounds occurring in minute amounts in complex mixtures, synergism and masking, synthesis, and the problems of developing control protocols that utilize synthetic compounds.

The third in a series of three seminars [*Science* **157**, 464 (1967); **160**, 445 (1968)] on new biochemical approaches to pest control, "Control of Insect Behavior by Natural Products," was held January 16-18, 1968 in Honolulu, Hawaii. These seminars were cosponsored by the National Science Foundation and the Japan Society for the Promotion of Science as part of the United States-Japan Cooperative Science Program. The recent efforts of twenty scientists to identify naturally occurring compounds that elicit chemosensory behavior and to describe their modes of action were reported in the context of the meeting theme—collaboration between biologists and chemists. Chemosensory behavior; electrophysiology; isolation, identification, and synthesis of new compounds; and applications to pest suppresion were the key areas discussed. (From: Wood, D. L., Silverstein, R. M., and Nakajima, M., 1969. *Science* **164**, 203. Copyright 1969 by the American Association for the Advancement of Science.) The papers presented at the third seminar have been revised and up-dated for publication in this volume.

The editors are deeply indebted to Kathleen Green for typing the finished copy; to Barbara Barr, Alan Cameron, and Caroline Wood for their devotion to editorial excellence; and to Celeste Green and Lewis Edson for redrafting many of the figures and formulas.

PHEROMONE RESEARCH WITH STORED-PRODUCT COLEOPTERA

Wendell E. Burkholder

Market Quality Research Division
U.S. Department of Agriculture
Department of Entomology
University of Wisconsin
Madison, Wisconsin

Table of Contents

Table of Contents--(Continued)

CONTROL OF INSECT BEHAVIOR

I. Introduction

Sex pheromones appear to play an essential role in the mating behavior of the stored-product Coleoptera. Information on how sex pheromones can be used to influence the behavior of these insects may well permit the development of safer and more effective control measures than are now available.

The earliest evidence of a stored-product beetle sex pheromone was for the yellow mealworm, Tenebrio molitor L. (Valentine, 1931). More recently Tschinkel et al. (1967) have reported that the pheromone is produced by both the female and male and acts as a weak attractant, and that its major role is sexual excitation of the male. A substance produced by the female Trogoderma granarium Everts attracts both females and males and was called a pheromone by Finger (Bar Ilan) et al. (1965). The males have been shown to produce a similar substance that attracts both sexes (Yinon and Shulov, 1967). Levinson and Bar Ilan (1967) described the function and properties of the assembling scent of this insect. A pheromone of the male bruchid, Acanthoscelides obtectus (Say) was reported by Hope et al. (1967). The biological function of this substance is unknown. Sex pheromones that influence the behavior of males are produced by the unmated female adults of the black carpet beetle, Attagenus megatoma (F.) (= A. piceus (Olivier)), Trogoderma inclusum LeC., and T. glabrum (Herbst) (Burkholder and Dicke, 1966). The first successful isolation, identification, and synthesis of a stored-product insect sex pheromone was accomplished with the black carpet beetle (Silverstein et al., 1967).

The present work is an examination of sex attraction in the black carpet beetle with notes on several other stored-product beetles.

II. Insect Rearing

The methods of rearing and handling the black carpet beetle have been described previously by Burkholder and Dicke (1966). The Trogoderma species were reared and handled in a similar manner except that the medium was improved during 1966 by adding to the dog food the following: dry milk, wheat germ and brewer's yeast for a weight ratio of 3:3:3:1. The cigarette beetles, Lasioderma serricorne (F.), were reared on whole wheat flour with approximately

3

5% brewer's yeast added. The cigarette beetles were sexed
as pupae by examination of the genitalia since adult sexual
differences are not easily determined.

III.   Three-Choice "Closed-System" Olfactometer

The initial discovery of sex pheromones in females of
the black carpet beetle, and T. inclusum and T. glabrum,
was made in a 3-choice olfactometer that consisted of a
modified glass desiccator (Burkholder and Dicke, 1966). The
insects, while in the chamber, were in an environment with
little air movement, which is similar to that often encountered by stored-product insects in nature. In this olfactometer the insects had the choice of either male or female
odor or no odor at all. In the presence of an attractive
odor, the male black carpet beetles exhibited the following
behavior: (1) a forward and upward extension of the antennae; (2) a "humping" behavior that was brought about by the
extension of the first pair of legs until they were nearly
straight, and the partial extension of the second pair,
which produced an angle of approximately 45° between the body
and the substrate; (3) a rapid zig-zag pattern of approach
to the attractant with intermittent stops, during which the
"humping" behavior occurred; and (4) copulatory attempts
with other test males. All of the first 3 responses were
necessary before a response to an odor was considered positive. Trogoderma males responded in a similar fashion
except for the "humping" behavior.

In tests with L. serricorne this olfactometer was
modified by adding Fluon (polytetrafluoroethylene dispersion) to the sides of the arena to keep the insects from
crawling up the sides (Radinovsky and Krantz, 1962). Males
and females were held separately from the time of adult
emergence. They were exposed in groups of 10 to the female
and male odors which were obtained by holding the insects
for 7 days in 5 dr shell vials with 12.7 mm paper discs on
the bottom. The paper discs were then transferred to the
chamber for the assay.

The data indicated that the males were attracted only
by the female odor (Table I). Little female response to
odors of either sex was observed.

Table I

Response of <u>Lasioderma</u> <u>serricorne</u> Males and Females to
Odors from Males and Females in a Desiccator Olfactometer

Sex of test insects	Average numbers[a] responding to		
	Female	Male	Control
Male	7.30	0.02	0.02
Female	0.35	0.50	0.52

[a] Averages of 8 replicates, each with 5 observations at
1 min intervals, with 10 insects per replicate.

IV. Multichoice Olfactometer with Air Flow

To determine whether or not males could differentiate
between several odors presented at the same time and at
greater distances, another olfactometer was developed. The
basic design was fan-shaped with the attractant choice available to the test insects at one or more of 5 points equi-
distant from the place of release (Fig. 1). After preliminary models of wood were built and tested, a unit was constructed of brass and glass. The 2 straight sides were 47
cm long, the curved side was 51 cm long, and the 4 dividers
were 12.7 cm long, all being 2.54 cm high and 6.35 mm thick.
The distance from the curved side to the apex was 44 cm and
the distance to the release point was 39 cm. The plate
glass top and bottom were 6.35 mm thick. There was a 1.27
cm hole in the top plate for insertion of the inverted
funnel for beetle release. The funnel was a modified Gooch
filter tube, 25 mm I.D. and 9 cm long. The funnel was short-
ened to reduce the depth of its widest part to 20 mm. When
the funnel was in the raised position, about a 5 mm opening
allowed free movement of the insects into the olfactometer.
A sheet of Whatman filter paper No. 1 covered the

lower plate glass and formed the floor of the chamber. The
filter paper and brass unit were sandwiched between the
two pieces of plate glass and sealed tightly by means of 10
clamps. The filter paper helped to form a gasket on the
bottom, and a 6.35 mm strip of masking tape provided a gasket under the top plate glass. A brass needle valve was
attached to the apex of the unit, and a threaded opening was
centered in each of the 5 divisions. Brass "Swagelok"
fittings with teflon ferrules for 6.35 mm tubing were attached. Each attractant-holding chamber was a 25 ml Erlenmeyer flask fitted with a 19/22 ground glass fitting bearing
an inlet tube extending into the flask and an outlet tube
for attachment to the olfactometer. The inlet tube was
attached by means of nylon "Swagelok" fitting to a Gelman
flowmeter.

The unit was attached to a 56 x 56 cm table with four
20 cm legs. This in turn rested on a movable wooden laboratory cart. Suspended 91 cm above the apex of the olfactometer was a 150 w soft-white light bulb. The apparatus
was placed in a room separate from other working and insect
culturing areas. Room ventilation was provided and attractants from the olfactometer were removed by means of a
vacuum system to avoid contamination of the room air. The
temperature usually ranged between 25° and 27°C. The relative humidity was regulated at 50 ± 5%.

Air flow through the flasks was regulated at 1/4 l/min.
Twenty-five test males were placed in the lowered inverted
funnel and allowed to settle for a few minutes before the
test. The air flow through the flasks was begun 30 sec
before release of the males. Observations were usually made
at 1 min intervals for 10 min.

A. Influence of Age of Female Attagenus megatoma on Male
    Response

A. megatoma male responses to females of various ages
were measured in this olfactometer. Ten newly-emerged
virgin females were placed on filter paper in Erlenmeyer
flasks. The flasks were capped with an aluminum foil covered
cork, placed in an incubator, and held for < 1, 2-3, 4-5,
and 6-7 days. Females held 24 hr or less from adult emergence were placed in the flasks 3 hr before the test. The
males used in this test were 6-7 days old. Two replicates
with different flasks and test males were run on separate
days.

6

CONTROL OF INSECT BEHAVIOR

The data indicate that the females were not attractive until at least 24 hr after emergence (Table II). Male
response was highest 6-7 days after female emergence, and
was nearly as high 4-5 days after emergence.

Table II

Response of Attagenus megatoma Males to Odors from
Females of Various Ages in a Multichoice Olfactometer

Age of	Numbers responding		
female	Replicate		
(days)	I	II	Average
< 1	0.1	0	0.05
2-3	2.6	1.5	2.05**
4-5	6.7	7.8	7.25**
6-7	8.8	8.3	8.55**
Control	0	0.1	0.05

** Values significant at the 1% level.

B. Influence of Mating of Attagenus megatoma on Male
    Response

As a preliminary to this test a series of petri dishes
was set up, each containing 1 male and 1 female of the
same age. The following age levels were used: < 1, 1-2,
2-3, 3-4, and 4-5 days following emergence. No copulation
was observed in the dishes at the first 3 age levels. Copulation occurred after 4 min at the 3-4 day age level and
between 1 and 3 min at the 4-5 day level. Ten pairs were
selected from the 4-5 day age level for further testing.

7

WENDELL E. BURKHOLDER

These insects remained together for 1/2-1 hr and were then
removed to separate 5 dr test vials. At the same time 10
unmated males and females of the same age were placed in
separate vials. These vials were fitted to the multichoice
olfactometer by inserting into the plastic lid 2 L-shaped
glass tubes which functioned as inlet and outlet ports. The
vials were tested in the same manner as the previous tests
after they had been set up 1 day and were retested after 7
days.

The data indicate that the mated female is somewhat
less attractive 1 day after mating and much less attractive
8 days after mating (Table III). The unmated females were
equally attractive at ages 5-6 and 12-13 days.

Table III

Response of Attagenus megatoma Males to Odors from
Unmated and Mated Males and Females in a Multichoice
Olfactometer

Insect	Average numbers[a] responding to				
age	Male		Female		Control
(days)	Unmated	Mated[b]	Unmated	Mated[b]	
5-6	0.6	0.4	11.1	8.1	0.1
12-13	0.3	0.7	11.1	4.1	0.3

[a] Averages of 10 observations at 1 min intervals (25 males).
[b] Mated at age 4-5 days.

C. Pheromones from Female Trogoderma parabile and Trogo-
    derma simplex

Females were held in 1 dr vials for 10 days and then
transferred to the olfactometer for assay.

The males of both species responded to female odors of
their respective species (Table IV).

8

CONTROL OF INSECT BEHAVIOR

Table IV

Response of Trogoderma parabile and Trogoderma simplex
Males to Female Odor of Their Repsective Species in a
Multichoice Olfactometer

Species	Control 1	Control 2	Female odor	Control 3	Control 4
	Average numbers[a] responding to				
T. parabile	1.03	0.53	9.17	0.67	0.97
T. simplex	0.20	0.73	17.70	0.57	0.37

[a] Averages of 3 replicates, each with 10 observations at 1
min intervals, with 25 insects per replicate.

V. Single-Choice Olfactometer

In order to conduct quantitative assay studies with
greater precision, a simple small closed-system olfactometer was developed (Fig. 2). A 5 mm x 6 cm glass rod with
one end flattened was inserted through a hole in the plastic
lid of a standard 5 dr opti-clear vial (28 x 55 mm). Paper
assay discs bearing the extract were affixed to the glass
rod by a 6-7 mm square piece of double-stick Scotch tape
and suspended 5-6 mm above the filter paper (24 mm disc
Whatman No. 2) floor of the chamber. Four test insects
were usually used in each chamber.

A. Quantitative Assay of the Attagenus megatoma Pheromone

A female extract was obtained by placing 4 virgin black
carpet beetle females less than 24 hr old in a 16 x 125 mm
screwcap test tube which was lined on one side with filter
paper. The females were held 4 days in an incubator at
26° ± 1°C, after which a stock extract was prepared by

9

adding 15 ml of benzene to the tube and macerating the beetles and filter paper. A 10 μl sample of the diluted benzene extract was placed on a 12.7 mm antibacterial assay disc and allowed to evaporate for 2 min. After 1 min the disc was turned over to permit evaporation from the other side. The disc was then inserted into the test chamber. One replicate consisted of 4 males confined to each of 6 chambers (5 different concentrations and a control). The males were observed at 1/2 min intervals for 10 min. Males that exhibited antennal extension, "humping" behavior, and rapid zig-zag movement were considered to have responded. These first 3 phases of the sexual response did not appear to be affected by the presence of other males. Copulatory attempts were not used as a basis for response in these tests as they were considered too dependent on other males or objects and also appeared to require a higher concentration of the pheromone.

Once the male had responded, the characteristic behavior usually persisted for the remainder of the test period. However, if a male that had responded settled down in a non-excited state, it was considered to have responded and therefore all counts were designated as "numbers of males responded". Those chambers containing 4 males that had responded were not observed further during the first test round. After the test the assay disc was removed and the lid of the chamber removed for ventilation. After 45 min rest the males were tested again with a fresh aliquot of the extract. The test was repeated on the second, third, and fourth days with 2 test rounds each day for a total of 8 tests, in which 32 insect responses were recorded for each replicate and treatment. The same stock extracts and males were used for these successive daily tests. A second and third replicate were started on successive days and tested in an identical manner; however, different extracts and test males were used for each replicate. Thus, a total of 96 insect responses was recorded at each treatment level. The tests were conducted at 26° ± 1°C and 50 ± 10% RH, under normal room light conditions, and during the afternoon hours when peak activity normally occurs.

A 50% response level after 3 min is estimated to occur at a concentration equivalent to 0.00032 of a female (Fig. 3).

B. Influence of Age of Male Attagenus megatoma on Male
    Response

    Males of 6 ages ranging from 1 to 20 days were exposed
to the female pheromone. The methods of preparing the
pheromone and conducting the test were the same as those
used in the quantitative assay described above. The responses of the males to the pheromone during a 5 min period were
noted at 1 min intervals. After 45 min the procedure was
repeated for each of 4 replicates.
    Males aged 1 day or less did not respond, while 97% of
the males aged 6-7 days responded (Table V).

Table V

Response of Attagenus megatoma Males of Various Ages to
a Benzene Extract Equivalent to 0.0004 of a Female

Age of males (days)	Per cent response at 5 min[a]
< 1	0
2-3	6
4-5	72
6-7	97
12-13	94
19-20	84

[a] Averages of 4 replicates, each with 4 males in a 5 dr
test chamber.

11

C. Interspecies Attraction with Attagenus megatoma,
    Trogoderma inclusum, and Trogoderma glabrum

For each species, 4 newly-emerged females were placed
in each of 3 shell vials (15 x 45 mm) with absorbent paper
discs on the bottom. The vials were held for 5 days after
which the paper discs were removed and attached to the glass
rod of the olfactometer. Ten vials of 5, 4-5 day old males
were set up for each species. The test was conducted by
inserting the glass rod with an attached paper disc successively into each of the test chambers for a maximum of 3
min. Exposure of males to the odors of females of the same
species was conducted last in the test series.
    None of the A. megatoma males responded to the T.
inclusum or T. glabrum female odor, but all responded to the
odor of female A. megatoma after only 9 sec exposure. None
of the T. inclusum males responded to the A. megatoma
females, but 98% responded to the T. glabrum female odor
within 3 min. All of the T. inclusum males responded to the
T. inclusum female odor in 1-3 sec. Similarly, none of the
T. glabrum males responded to the A. megatoma female odor,
but 100% responded to the T. inclusum odor after 20 sec
exposure. However, nearly 3 min were required for a 100%
response to the female T. glabrum odor.

D. Response of Attagenus megatoma Males after Removal of
    Distal Portions of the Antennae

Males less than 24 hr after emergence were chilled for
a few minutes in a refrigerator, and part of the antennae,
which included the clubs (segments 9-11), were removed between segments 2 and 8 with forceps. Three vials each
contained 1 male with the antennae thus removed, and an
intact male in a fourth vial served as a control. This
series was replicated 9 times. The males were 4-7 days old
when exposed to the pheromone. The female odor was obtained
by holding 4 newly-emerged females on a paper disc in each
of 3, 15 x 45 mm shell vials for 4 days. The glass rod with
the pheromone-impregnated disc attached was placed into each
of the 4, 5 dr vial olfactometers for a maximum period of
5 min.
    None of the 27 males with the antennae removed responded
to the female odor, while the 9 control males all responded
in 1-3 sec.

In order to assess mating success, each male was confined with a female in a 2 dr vial. Eight of the 9 control
males copulated with a female within a few seconds. Of the
27 males with part of the antennae removed, only 2 were
observed to copulate with a female within the 1/2 hr observation period; 1 copulated after 5 min and the other after
10 min.

E. Response of Attagenus megatoma Males to the Synthetic
    Pheromone and Its Isomers

The isolation, identification and synthesis of the
black carpet beetle pheromone have been described by
Silverstein et al. (1967). Megatomoic acid (trans-3, cis-5-
tetradecadienoic acid) and the other 3 isomers were synthesized and tested.
    The results of the bioassay show that the 3 isomers were
less active than the trans-3, cis-5 isomer isolated from the
female beetle (Table VI). The apparent activity shown by
the other isomers remains in doubt because of contamination
with small amounts of the trans-3, cis-5 isomer. This contamination was unavoidable because a small amount of iso-
merization occurs during alkaline hydrolysis of the methyl
esters to obtain the carboxylic acid.

Table VI

Response of Attagenus megatoma Males to Four
isomers of the Female Pheromone

Pheromone isomer	Numbers responding[a] to the following concentrations			
	0.1 µg	0.01 µg	0.001 µg	0.0001 µg
trans-3, cis-5				
(megatomoic acid)	16	16	11	2
cis-3, cis-5	16	16	5	0
cis-3, trans-5	16	5	1	0
trans-3, trans-5	12	3	0	0

[a] Sixteen males per test.

13

VI. Y-Tube Olfactometer

In studies of the attractants of the cigarette beetle,
L. serricorne, a modified Y-tube olfactometer was developed
(Fig. 4). This olfactometer was one of the many modifica-
tions of the Y-tube olfactometer developed by McIndoo (1926,
1933), and based on the principle first developed by Barrows
(1907). The arms were bent down at a 90° angle and inserted
into 25 ml filter flask traps. Air passed through flowme-
ters into vials containing the odors, then into the filter
flask, and finally into the olfactometer. Two olfactometers
were set up in parallel. Ten 9 day old females were released
into one and 10, 9 day old males into the other. Both sexes
were given the choice between odors from the 9 day old males
or females. Fluon was used to prevent the beetles from
climbing out of the flasks.
     The males exhibited a strong response to the female
odor (Table VII).

Table VII

Response of Lasioderma serricorne to Male or Female
Odors in a Y-Tube Olfactometer

Sex of test insects	Average numbers[a] responding within 15 min to odor of	
	Females	Males
Male	8.50	0.50
Female	1.25	2.50

[a] Averages of 4 replicates with 10 insects per replicate.

CONTROL OF INSECT BEHAVIOR

VII. Field Trap

A field trap was constructed from an 18 x 18 x 1.5 cm piece of Styrofoam. A hole was cut in the center of the Styrofoam block to receive a 9 cm plastic petri dish, and the upper surface was sloped to facilitate insect access to the dish. In some tests, another 9 cm plastic petri dish with 4 holes cut in the side was inverted over the inserted dish and fastened to it with tape. A 9 cm filter paper impregnated with the synthetic A. megatoma attractant was placed on the bottom of the dish. The insects crawled up the Styrofoam block, entered the trap through the holes in the upper dish, and fell into the lower dish.

In laboratory tests males were released in a 56 x 66 cm tray. Field tests were conducted in an empty poultry building that served as a warehouse. Males were released at one end of a 61 x 183 cm enclosure. In laboratory and field trials, about 80 and 60%, respectively, of the released beetles were trapped (Tables VIII, IX).

Table VIII

Response of Attagenus megatoma Males to trans-3, cis-5-
tetradecadienoic Acid Contained in Field Traps
and Exposed in the Laboratory

Treatment	Average numbers responding[a] after					
	5 min	15 min	30 min	1 hr	2 hr	18 hr
Synthetic attractant (0.01 mg[b])	7.7	14.3	21.3	29.3	34.3	39.3
Control	0.3	0.3	1.0	1.0	1.0	1.6

[a] Averages of 3 replicates, each with 50, 8-9 day old males.
[b] Equivalent to 20 females.

Table IX

Response of Attagenus megatoma Males to trans-3, cis-5-
tetradecadienoic Acid Contained in Field Traps
and Exposed in a Warehouse

Treatment	Per cent response[a] after		
	3 hr	6.5 hr	22.5 hr
Synthetic attractant (0.25 mg)	28.7	45.0	50.0
Control	2.5	7.5	7.5

[a] Eighty males released.

VIII. Discussion

There is increasing evidence that stored-product
beetles rely on pheromones for communication between the
sexes. A. megatoma females apparently produce 1 compound
that is an attractant to the male. Among the Trogoderma
species a complex of chemicals is indicated. Isolation
studies by R. M. Silverstein and J. O. Rodin (Stanford
Research Institute) indicate that there are 4 active compounds produced by T. inclusum females that elicit responses
from the males. A similar pattern is indicated for T.
glabrum. It appears that many of the Trogoderma species
possess sex attractants that are not species specific. Multiple-compound pheromone complexes may occur in patterns
such that for one species one compound is the primary pheromone and the others are of trivial or secondary importance,
while for another species one of these secondary compounds
may be the primary pheromone.

Pheromones are probably essential to the survival of
stored-product beetles, particularly in low-density population situations. However, under high-density conditions,
pheromone production may be so high that adaptation of sensory cells on the male antenna occurs. This mechanism could

control population increase by decreasing mating success.
Stored-product insects are ideal subjects for pheromone
studies because they can be cultured and used for assays
throughout the year in the laboratory. Rearing the insects
is usually not difficult, and large numbers can be supplied
to a cooperating chemist for study. However, considerable
manpower is necessary to isolate, sex and prepare for de-
livery to the chemist the large number of insects required.
The bioassays can be accomplished in several ways and are
simple enough that the chemist can readily monitor his chromatographic and chemical isolation steps by means of the
insect response. Test males are very susceptible to contamination by the female sex pheromones prior to testing,
and special precautions must be taken to avoid this. If
females emerge in a culture jar, male pupae or adults in
the same jar may absorb the pheromone and become attractive
to other males. The most important consideration in studies of stored-product insect pheromones is that the bioassay conditions closely approximate their natural habitat.
Then the results of laboratory tests may be directly used
for field application when the chemical identity of the
pheromone becomes known.

Acknowledgments

This paper has been approved for publication by the
Wisconsin Agricultural Experiment Station, Madison. The
author is indebted to Mr. J. Gorman, Agricultural Research
Technician of this laboratory, for rearing the insects used
in this study; to Professor R. J. Dicke, Department of
Entomology, University of Wisconsin, for his cooperation and
advice; and to Dr. L. S. Henderson and Mr. D. P. Childs for
suggesting the study conducted with the cigarette beetle.
Fluon is a product of Imperial Chemical Industries.
Mention of a propietary product does not necessarily imply
its endorsement by U.S.D.A.

IX. References

Barrows, W. M. (1907). J. Exp. Zool. 4, 515.
Burkholder, W. E., and Dicke, R. J. (1966). J. Econ.
    Entomol. 59, 540.
Finger (Bar Ilan), A., Stanic, V., and Shulov, A. (1965).
    Rivista Di Parassitologia 26, 27.
Hope, J. A., Horler, D. F., and Rowlands, D. G. (1967).
    J. Stored Prod. Res. 3, 387.
Levinson, H. Z., and Bar Ilan, A. R. (1967). Rivista Di
    Parassitologia 28, 27.
McIndoo, N. E. (1926). J. Econ. Entomol. 19, 545.
McIndoo, N. E. (1933). J. Agric. Res. 46, 607.
Radinovsky, S., and Krantz, G. W. (1962). J. Econ. Entomol.
    55, 815.
Silverstein, R. M., Rodin, J. O., Burkholder, W. E., and
    Gorman, J. E. (1967). Science 157, 85.
Tschinkel, W., Willson, C., and Bern, H. (1967). J. Exp.
    Zool. 164, 81.
Valentine, J. M. (1931). J. Exp. Zool. 58, 165.
Yinon, V., and Shulov, A. (1967). J. Stored Prod. Res. 3,
    251.

Figure 1.--Multichoice olfactometer.

Figure 2.--Single choice olfactometer.

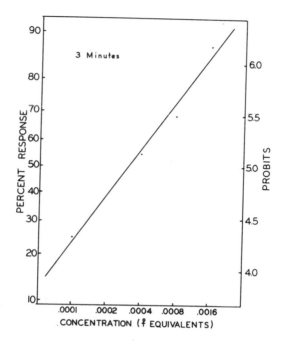

Figure 3.--Probit response of <u>Attagenus</u> <u>megatoma</u> males to
the female pheromone.

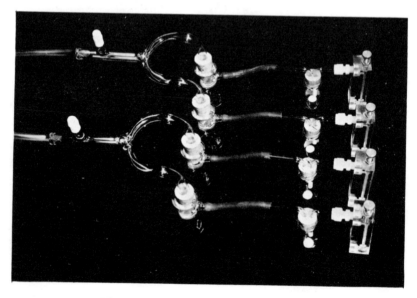

Figure 4.--Y-tube olfactometer.

SOME GENERAL CONSIDERATIONS OF INSECTS' RESPONSES TO
THE CHEMICALS IN FOOD PLANTS

V. G. Dethier

Department of Biology
Princeton University

As is true of all animals, the response of an insect
to its chemical environment assumes two forms: delayed reactions and immediate reactions. Delayed reactions include
symptoms of toxicity, growth, hormonal changes, and caste or
reproductive differentiation. In some instances the chemicals gain entry to the body by way of the cuticle or respiratory system. This is the case with contact insecticides.
In other instances chemicals enter via the digestive tract.
This is true of insecticides that must be ingested to be
effective and of all nutrients and special growth substances.
The time-course of post-ingestive responses varies over a
very wide range from the comparatively slow processes of
growth and elaboration of reproductive products to the
rapid onset of such behavioral changes as the transition
from hunger to satiation.

Immediate responses comprise the many behavioral mani-
festations of the interaction of chemicals with exterocep-
tive sense organs. Overt behavioral responses to stimulation of these senses include the initiation or termination
of locomotion, feeding, courtship, copulation, oviposition,
grooming, the defense of territory, aggressive actions,
defense actions, nest construction, and various aspects of
parental care. Unfortunately, a very loose vocabulary came
to be employed in categorizing chemicals in terms of the
behavioral effects that they elicited from insects. An
attempt to inject some meaningful order into this nomencla-
ture was made a number of years ago (Dethier et. al., 1960).
(The original paper should be consulted for an understanding of the underlying rationale.) For present purposes it
is sufficient to recall that chemicals eliciting immediate
behavioral responses in insects may be categorized as:
attractants (chemicals that cause insects to make oriented
movements toward the source), arrestants (chemicals that
cause insects to aggregate), stimulants (chemicals that
elicit feeding (phagostimulant), oviposition, etc.), repellents (chemicals that cause insects to make oriented
movements away from the source), deterrents (chemicals that
inhibit feeding or oviposition). Chemicals of plant origin
may fit into any of these categories. Their action is most
commonly associated with some link in the chain of behavior
leading to ingestion or oviposition.

The senses primarily involved in the detection of
plant chemicals are the olfactory and gustatory. To under-
stand thoroughly the behavioral relation between insects

and their host plants it is necessary to have a precise
identification of the stimulus and a comprehension of the
action spectrum (specificity or non-specificity) of the
chemoreceptors as well as of their coding characteristics.
The most trying handicap facing these endeavors today is the
almost total lack of knowledge of plant chemistry as it
relates to insect behavior. If there were lists of the
chemicals present in a particular host plant of a particular insect, it would be possible to select for bioassay
those that appeared to be the most likely candidate stimuli.
In fact, whenever one is faced with the task of ascertain-
ing which compound or mixtures of compounds in a given host
plant act as attractants, arrestants, stimulants, repellents, or deterrents, the existing vast chemical literature
is of limited value, and the investigator must undertake
the isolation and characterization himself.

Once the identities of the stimuli are known, and they
have been characterized as to the form of behavior that
they elicit, the next step is to discover to what extent
the behavior is dictated by the filtering capacities of the
sense organs (that is, their specificity), how great a con-
tribution is made by the command and decision-making inter-
neurons of the central nervous system, how much is genetically built into the organism, and how great a role is
played by experience. At the moment we are less handicapped
by lack of information about insect chemoreceptors than by
ignorance of plant chemistry because studies of sensory
physiology have advanced rapidly during the past two decades.
It must be admitted, however, that not much of the knowledge
has been derived from investigations with phytophagous
species. Most of the basic techniques have been perfected
in experiments dealing with flies of the genera Phormia and
Calliphora, butterflies of the genus Vanessa, and various
silkworm moths. Only recently have the techniques been
applied to the study of phytophagous species, and even then,
almost exclusively to lepidopterous larvae (see Hanson in
this symposium for a summary). Unfortunately when plant
chemistry has been studied in detail, it has usually been
in connection with plants of economic importance; and
through some irony of nature the pests of these plants are
generally small and do not lend themselves readily to
neurological and physiological analyses.

Despite the drawbacks a cautious extrapolation from
the data obtained with "laboratory" insects is informative

23

especially when coupled with the existing data on lepidopterous larvae. The general view obtained gives the impres-
sion of a considerable degree of receptor specificity. In
the blowfly _Phormia_ _regina_ (Meigen), for example, there are
receptors narrowly specific for certain carbohydrates, some
specific for salts, some restricted to sensitivity to water,
and some on the tarsi probably specific to acids. A thirsty
fly accepts water and dilute sodium chloride as a consequence of impulses generated by the water receptor in one
case and by low frequency discharge of the salt receptor in
the other case. A water satiated fly accepts sugar solutions as a consequence of impulses generated by the sugar
receptor. Thus it may be said that water, dilute solutions
of sodium chloride, and some carbohydrates are feeding
stimulants for the fly. It is clear from these results
that three entirely different classes of compounds can
elicit the same kind of behavior, that is, ingestion. If
one were analyzing a food in order to discover which components were effective stimuli, or wished to employ known
stimuli as a starting point for synthesizing superoptimal
stimuli, confusion would ensue unless it was clearly recognized that the three classes do not act via a single receptor; that is to say, a feeding stimulant in this case is
not something that possesses in common certain features of
water, salt, and sugar. The experiments with _Phormia_ reveal,
furthermore, that a chemical which may act as a feeding
stimulant under one set of circumstances (e.g., dilute salt
when a fly is water-deprived) is not an effective stimulant
under other circumstances (e.g., when the fly is water-
satiated).

The case just described probably reflects the situation
in lepidopterous larvae also. These insects possess specific receptors for water, glucose, fructose, inositol, salts,
and some kinds of deterrents (cf. Ishikawa, 1963; Ishikawa
and Hirao, 1963, 1966; Schoonhoven and Dethier, 1966).
Impulses generated by the first three receptors mediate
acceptance; those generated by the remaining receptors
mediate rejection. Thus, natural plant products that are
adequate stimuli for the first three receptors could be
arrestants and most certainly are feeding stimulants; those
that are adequate stimuli for the other two receptors are
deterrents. The outcome of any feeding situation would be
determined by the ratio of input of the two classes of
receptors (cf. Schoonhoven and Dethier, 1966). The exis-

tence of gustatory receptors of even greater specificity is indicated by Schoonhoven's (1967a) discovery that there are in each maxilla of larvae of Pieris brassicae (L.) two receptors that are especially sensitive to mustard oil glycosides, the feeding stimulant found in the host plants.

At the present level of knowledge the response characteristics of the olfactory system appear different than those of the gustatory system; however, further work may invalidate the distinction. Whereas taste receptors tend to be individually specific to restricted classes of compounds, olfactory receptors may be either specific or nonspecific. Schneider and his co-workers (1965), who have conducted extensive experimentation with the sex attractants of silkworm moths and with odors to which honeybees are responsive, recognize two categories, specialists and generalists. The specialists are receptors that respond to a restricted group of substances, not necessarily related chemically. For example, there is a "grass receptor" in the locust and a "carrion receptor" in Calliphora. The generalists are receptors which respond to a wide variety of odors with action spectra that overlap; however, no two spectra are identical.

The antennal olfactory receptors of Manduca sexta (Johan.) respond to a wide variety of odors and resemble "generalists" (Schoonhoven and Dethier, 1966). In the absence of stimulation they are spontaneously active. Stimulation may increase or decrease the rate of firing depending on the identity of the stimulus and the particular receptor. Thus, a given odor may increase the rate of firing of one receptor and decrease the rate of firing of another while a different odor may have exactly the opposite effect. The receptors have overlapping but not congruent action spectra. As a consequence, divers odorous compounds, by their differential effects on the several receptors, cause different compound patterns of activity to be generated when all receptor input is viewed as a whole (cf. Dethier, 1967). In short, the gustatory and olfactory systems of lepidopterous larvae can transmit information by generating compound patterns of action potentials, by filtering stimuli by means of highly specific receptors (e.g., the glucose receptor), or by combining both methods.

It is primarily the information transmitted by either or both of these methods of coding which determines whether a plant will be accepted or rejected and to what degree.

At the moment the experimenter is less adept at translating these codes into behavioral terms than are the caterpillars. Hamamura (this symposium) has described how different compounds trigger different steps in the feeding chain of response of Bombyx mori (L.). It would be illuminating to be able to match this work with electrophysiological analyses and thus learn more about the neural bases of the behavior patterns. Ishikawa (1963) and Ishikawa and Hirao (1963, 1966) have taken some steps in this direction.

To complete the analysis of feeding behavior at this level it must be noted that other sensory modalities may play a supporting role. Hanson (this symposium) has described maxillary tactile receptors that may be involved in detecting textural differences in leaves. Schoonhoven (1967b) and Dethier and Schoonhoven (1968) have found in the antennae of lepidopterous larvae temperature receptors of a sensitivity sufficient to enable larvae to detect temperature differences in transpiring and non-transpiring leaves. Furthermore, temperature changes arising from the evaporation of leaf fluids when a caterpillar takes a single bite are detectable by the antennal receptors. By means of these receptors and additional hygroreceptors larvae have at least the potentiality for monitoring evaporation as such. Detection of evaporation and of transpiration conceivably could inform the larvae of the state of water balance of the plant. Thus far it has not been possible to devise a behavioral experiment to test this hypothesis.

With all of the sensory input available to a larva the extent to which the decision-making is constrained by the filtering properties of receptors or by central decoding of multiple overlapping patterns of input remains a mystery. Regardless of the level in the nervous system at which the decision occurs, it is genetically based; that is, there is a species-characteristic innate spectrum of feeding preferences. Within this limit, however, changes in preference due to experience are possible, as Jermy et al. (1968) have demonstrated with the tobacco hornworm (M. sexta) and the corn earworm (Heliothis zea (Boddie)).

The hornworm is an oligophagous insect feeding on solanaceous plants in nature. It can feed and develop on other plants when deprived of its maxillary taste receptors (Waldbauer, 1962). When larvae that have been fed since hatching on a diet lacking plant material are presented with

a choice of Lycopersicon, Solanum, or Nicotiana, no significant preferences are observed. When larvae that have been
raised exclusively on one or other of the three plants are
tested for preference, the Lycopersicon- and Solanum-fed
individuals exhibit a marked preference for the plant upon
which they had been feeding. No preference for non-host
plants could be induced by feeding larvae on a diet mixed
with leaves of the plants in question even though the mixtures were accepted as food.

Experiments with the polyphagous H. zea showed that
preferences could be induced in this species also. Feeding
for as short a time as 24 hours was sufficient to modify
preference. Furthermore, preferences induced in one instar
survived through one instar and two larval molts; that is,
diet-reared larvae fed on a plant for 24 hours, then returned to diet for a molt, an instar, and a second molt,
then given a preference test, retained their preference for
the inducing plant.

It is clear that feeding an oligophagous or polyphagous
caterpillar on a particular plant can induce a preference
for that plant provided it is a normal host plant. The
induced preference is specific for the inducing plant and is
not merely a change in the insect's general threshold of
food acceptability. This equality of consumption could
occur as a consequence of similar amounts of a common
phagostimulant in the three plants or of three different
taste qualities equally preferred. If the former, then
induction of increased feeding on one plant should also
cause increased ingestion of the others. An increased
consumption of one plant is associated with a relative
decrease of the others. This finding indicates, therefore,
that food selection in at least these two phytophagous
insects is not governed only by the quantities of phagostimulants and deterrents but that it involves the differentiation of complex qualitative information which in turn
enables the larvae to discriminate among different host
plants (and perhaps strains) as well as between host and
non-host plants.

References

Dethier, V. G. (1967). *In* "Handbook of Physiology", Sec. 6, 1, 79. (Editor, Code, C. F.) Amer. Physiol. Soc., Washington, D.C.

Dethier, V. G., Barton Browne, L., and Smith, C. N. (1960). *J. Econ. Ent.* 53, 134.

Dethier, V. G., and Schoonhoven, L. M. (1968). *J. Insect Physiol.* 14, 1049.

Hamamura, Y. (1970). (This symposium).

Hanson, F. E. (1970). (This symposium).

Ishikawa, S. (1963). *J. Cell. Comp. Physiol.* 61, 99.

Ishikawa, S., and Hirao, T. (1963). *Bull. Seri. Exp. Sta.* (Tokyo) 18, 297.

Ishikawa, S., and Hirao, T. (1966). *Bull. Seri. Exp. Sta.* (Tokyo) 20, 291.

Jermy, T., Hanson, F. E., and Dethier, V. G. (1968). *Ent. Exp. Appl.* 11, 211.

Schneider, D. (1965). *Symp. Soc. Exp. Biol.* 20, 273.

Schoonhoven, L. M. (1967a). *Koninkl. Neederl. Akademie van Wetenschappen* - Amsterdam *Proc.*, Series C. 70, 556.

Schoonhoven, L. M. (1967b). *J. Insect Physiol.* 13, 821.

Schoonhoven, L. M., and Dethier, V. G. (1966). *Arch. Néerl. Zool.* 16, 497.

Waldbauer, G. P. (1962). *Ent. Exp. Appl.* 5, 147.

PHEROMONES OF THE HONEY BEE, Apis mellifera L.

Norman E. Gary

Department of Entomology
University of California
Davis, California

Table of Contents

I. Communication by Pheromones in Social Insects

Social insects have extremely complex communication systems. The operation of these systems is contingent upon a variety of mechanisms of information transfer between colony members. Such communication facilitates the coordination of a multitude of behavioral interactions between thousands of individuals engaged in such activities as nest construction, brood rearing, and colony defense. Each member of an insect society must receive endless "messages" concerning the behavioral status of other colony members, and, reciprocally, must send similar "messages". The degree of social organization is correlated with the frequency and complexity of these "messages". The exchange of such "messages" may be regarded as communication, defined here as the intra-species transfer of stimuli that elicit behavioral or physiological responses in receptive individuals. Although all stimulus modes (auditory, optic, tactual, chemical, etc.) are used in information transfer or communication, pheromones are used as a primary mode of communication in social insects. A realistic appreciation of communication by pheromones involves the realization that information transfer is usually achieved by the use of pheromones in conjunction with other stimulus modes.

II. Dynamics of the Honey Bee Colony

A typical honey bee colony contains 20,000 to 60,000 bees, composed of three castes: (1) a single queen, the only fully-developed female in the colony, and normally the source of eggs from which all other colony members develop; (2) several hundred drones, or males, (absent during the winter) that serve the sole function of mating with the queen; and (3) thousands of workers, underdeveloped females that perform the activities associated with colony life, for example, brood care, colony defense, and food gathering and processing.

In nature, a honey bee colony normally occupies an enclosed nest space, such as a hollow tree containing one to several cubic feet of available nest space. Normally, the internal nest environment is quite dark. This tends to minimize or eliminate visual stimuli as a form of communication, and appears to place great selective pressure on other mechanisms of communication, especially the pheromone system. Furthermore, the nest environment is controlled

CONTROL OF INSECT BEHAVIOR

to some extent by the bees, in terms of rate of air exchange, humidity, and especially temperature ($94°F \pm 2°$ in
brood nest). Such a controlled environment in the enclosed
nest theoretically improves the efficiency of communication
by pheromones. For example, chemical messages, or signals,
can be released and diffused quickly and uniformly by active
ventilation (fanning). Thus, virtually all of the population is exposed quickly to pheromones released inside the
nest. Within a few seconds the pheromones can be dissipated
from the nest. The exceedingly high population density,
coupled with the ability of bees to reinforce or possibly
neutralize circulating chemicals by adding additional pheromones, appear to enhance the efficiency of the honey bee
pheromone communication system.

Honey bees have been utilized extensively in pheromone
research (see review by Butler, 1967a) because they are
ubiquitous and available in large populations that can be
conveniently maintained and studied in glass-walled observation hives. Furthermore, honey bees have a rich behavioral "repertoire" and many exocrine glands that produce
complex secretions that produce a large number of pheromones
(Fig. 1).

III. Intra-colony Effects of Queen Pheromones

Some of the most interesting and potentially useful
pheromones of honey bees are produced by the queen. These
pheromones appear to regulate worker-queen relationships
to a large degree. Workers seem to have an "awareness"
(expressed as normal behavior) of the presence of their
queen at all times (Butler, 1954a, 1954b, 1960a). Several
significant events occur soon after a colony loses its
queen. First, within minutes after queen removal from the
colony, bees respond by fanning and becoming generally
agitated. Usually within 24-48 hours, a few brood cells
containing worker larvae less than three days old are con-
verted into queen cells. The presence of a normal queen,
or queen cells, tends to inhibit oogenesis in worker ovaries.
In the absence of these inhibitory pheromones from the
queen, worker ovaries become functional to the extent that
small numbers of unfertilized eggs may be produced and laid.
Before the term "pheromone" was proposed, Butler (1954a)
coined the name "queen substance" to describe what appeared
to be a single inhibitory chemical produced by the queen.
Since that time several workers (Butler, 1954b; Butler and

Gibbons, 1958; Erp, 1960; de Groot and Voogd, 1954; Pain,
1961b; Pain and Barbier, 1960; Pain et al., 1962; Verheijen-
Voogd, 1959) have demonstrated that chemical extracts from
the queen's body are biologically active when exposed to
small groups of confined worker bees. In one bioassay
(Butler and Gibbons, 1958), workers were given young larvae
in honeycomb cells, and the number of cells subsequently
modified into queen cells was used as the criterion of
inhibitory activity. Other bioassays were based on the degree of ovary development of confined bees, as shown in
Table I. Nutritional factors apparently also influence ovary
development; consequently, results in these bioassays were
frequently quite variable, leading initially to some confusion in the literature. Laboratory bioassays utilizing
queen cell development were equally variable. Some of the
problem was resolved in later experiments when it became
apparent that multiple pheromones were involved (Butler,
1961; Pain, 1961a). These bioassays enabled Butler and
Simpson (1958) and others to determine that the mandibular
glands are a primary source of queen substance. It was
postulated that the mandibular gland secretion is spread
over the queen's body during grooming activities, and spread
within the colony by sharing regurgitated food (Butler,
1956).

In preliminary experiments, Voogd (1955) concluded
that the inhibitory substance had the characteristics of a
fatty acid. This clue was useful to the research teams that
ultimately succeeded in isolating (Butler et al., 1959;
Barbier and Pain, 1960), identifying (Callow and Johnston,
1960; Barbier and Lederer, 1960), and synthesizing (Callow
and Johnston, 1960; Barbier et al., 1960; Barbier and
Hügel, 1961; Butler et al., 1961) queen substance. Its
structure is 9-oxodec-trans-2-enoic acid.

$$CH_3-CO(CH_2)_5CH=CH-COOH$$

This pheromone, when complemented by other unidenti-
fied olfactory queen pheromones, is biologically active,
but less active than pheromones from a living queen (Butler
and Fairey, 1963). Probably some components of the pheromone system are still absent from the extracts. Highly
volatile or labile chemicals may be involved, and these
compounds are difficult to manipulate experimentally. It
is clear that all of the inhibitory pheromones of the
queen do not originate in the mandibular gland. When the

Table I

Effect of Live Queens, Live Queen Scent, and
9-oxodec-trans-2-enoic Acid, Separately and Together, on Ovary Development[a]

	A	B	C	D	E
	Control (no queen pheromones)	Access to live mated queen	9-oxodec-enoic acid only	Scent from live mated queen only	9-oxodecenoic acid + scent from live mated queen
Mean % bees per cage showing ovary development	76.7	7.7	36.7	60.0	31.3
S.E. of mean	2.15	2.33	2.50	2.79	2.60

a Adapted from Butler and Fairey (1963). The only difference not highly significant (P<0.001) is that between treatments C and E. Twelve cages, containing 40 workers less than 10 days old, were used for each treatment. At least 25 bees from each treatment were examined for ovary development.

mandibular glands were extirpated from living queens (Gary, 1961a) that were returned to their respective colonies, partial to complete inhibition of queen rearing activities (cell construction) was evident in normal colonies containing approximately 20,000 to 40,000 bees (Gary and Morse, 1962). Total inhibition of queen rearing and ovary development was caused by queens without mandibular glands, when tested in colonies with smaller populations on the order of 5,000 to 10,000 bees (Gary and Morse, 1962; Velthuis and Es, 1964). Generalized dermal secretions, as well as the secretion of Koschevnikov gland (Bulter and Simpson, 1965), and the subepidermal glands (Renner and Bauman, 1964), are other possible sources of pheromones that inhibit ovary development and queenless behavior in worker bees.

Velthuis et al. (1965) were able to demonstrate that laying workers and/or extracts of laying workers elicited retinue behavior and inhibited ovarial development in caged young bees under experimental conditions. This discovery becomes even more intriguing, when coupled with the report by Pain et al. (1967) that queen substance was not detected in laying workers. Apparently laying workers produce enough of the queen pheromones to mimic queens in some respects.

The mode of action of queen substance is not well understood. Most of the research is difficult to interpret because of the inherent complexities of the response. Also there are conflicting data, possibly caused by testing heterogeneous mixtures of pheromones in a variety of bio-assay conditions, while using bees of variable genetic composition. The injection of synthetic queen substance into workers does not inhibit queen rearing (Butler and Fairey, 1963). Several workers have suggested that this compound expresses its effect olfactorily, and subsequently affects the endocrine system. Gast (1967) described two growth phases in the development of the corpora allata in workers. In the first phase, development was inhibited by queen substance; in the second phase, development occurs only when a queen is present.

Ingested queen substance may also induce significant effects. Johnston et al. (1965) have shown that queen substance is metabolized very rapidly in the gut of worker bees. They suggested that the conversion of queen substance was sufficiently rapid to account for the expression of queenless behavior so quickly after bees are separated from their queen.

Queen pheromones appear to attract workers to the
queen where they are stimulated to feed and groom her.
Figure 2 illustrates the "retinue" behavior of workers as
they tend to encircle the queen, touch her with their
antennae, and lick the queen's body surfaces. The release
of feeding and grooming behavioral patterns in workers
possibly may be caused (a) by detecting an increase in concentration of attractant pheromones as workers come near
the queen; (b) by sensing additional pheromones that are
not involved in attraction per se; (c) by detecting essentially nonvolatile pheromones on the queen's body by antennal contact; and (d) by responding to a combination of
chemical, tactile, and perhaps sound stimuli. Most of the
"attractant" pheromones are produced in the mandibular
glands; living queens with extirpated mandibular glands
lose most (approximately 85%) of their attraction for
workers (Gary, 1961b; Morse and Gary, 1963; Zmarlicki and
Morse, 1964). Some of the attractive substances are secreted by the Koschevnikov gland (Butler and Simpson,
1965). The attraction of workers to queens can be tested
by determining the numbers of worker bees that cling to the
walls of cages containing queens, as shown in Fig. 3.
Occasionally, normal grooming and feeding of queens
becomes disrupted by other intensive activities that appear
initially to be a form of hyperattraction. Worker bees
crowd tightly around the queen. The temperature rises
sharply, and, in all probability, the queen and workers
release alarm pheromones while in this stress state. In
any event, aggressive behavior toward the queen results
and she is usually stung and killed by the workers. It is
not clear whether the aggressive behavior is a primary cause
or a secondary effect in this behavior, called "balling".
It is clear, however, that the clustering, aggressive,
balling response can be induced frequently by introducing
objects, such as filter papers, or other bees, coated with
mandibular gland pheromones (Gary, 1961c). This behavior,
apparently regulated by pheromone quality and/or concentration, may function as a rejection mechanism for abnormal
queens, e.g., queens producing or releasing pheromones
erratically or somewhat out of phase with colony requirements.
Although pheromones, by definition, elicit intra-
specific responses, the same compounds sometimes elicit
inter-specific responses. Butler (1966b) has reported that

pheromones from queens of Apis cerana, A. florea Odontotermes sp. (termite), and Formica fusca (ant), apparently inhibit queen rearing and ovary development in workers of A. mellifera. It should be remembered, however, that the fundamental communicative value of these "shared pheromones" is regulated by other variables in addition to the qualitative chemical differences, e.g., the concentration of chemical released, the behavioral context in which the pheromones are released, and the distribution variables (temporal and spatial) involved in the release system.

IV. Mating Attractant Pheromones

The phenomenon of conveying different pheromonal information, by virtue of liberating identical chemicals in a different context (Wilson, 1965), seems to be evident in the honey bee. For example, queen substance, an inhibitor of ovary development and queen cell construction within the hive, functions as a mating attractant outside the colony. Queens mate only while flying. Gary (1962) found that great numbers of flying drones were attracted to tethered virgin queens, extracts of queens, or synthetic queen substance. A typical cluster or swarm of drones hovering beneath a small cylindrical cage containing a virgin queen is illustrated in Fig. 4.

The extremely low vapor pressure of queen substance initially aroused skepticism that this compound could function as a sex attractant. However, several workers (e.g., Pain and Ruttner, 1963; Butler and Fairey, 1964) confirmed that queen substance is indeed the primary bioactive compound responsible for attracting drones to queens on mating flights.

There is a very fundamental question that has not been answered satisfactorily. . . to what extent is queen substance complemented by additional compounds? Gary (1962) reported that queen substance is only the primary compound in the queen sex attractant pheromone. He reported that the mandibular gland secretion loses most of its attractiveness to drones after fractionation, that is, no single fraction is very attractive in quantities approximating a normal queen equivalent. However, reconstituted mandibular gland secretion, made by combining the various fractions, restores most, if not all, of the attractive qualities and stimulates the formation of very large drone swarms. In fact, the large swarms of drones are quite impossible to count visually (must be photographed). Further support of the multiple-compound attractant hypothesis is found in the following observation (Gary, unpublished data): living virgin queens approximately 1-2 weeks old are significantly more attractive to flying drones than a wide range (10 µg to 10 mg) of synthetic queen substance displayed in similar cages. Attraction is measured by the numbers of drones hovering in clusters under the queen, and the duration of such clusters before they break away and reform. Virgin queens of the age used above normally contain a mean of approximately 130 µg of queen substance (Butler and Paton, 1962). This finding agrees well with the data of Pain et al. (1967) who detected a mean of 108 µg in queens that were assayed individually. However, there are no available estimates on the actual rate of release of queen substance by living queens.

Renner and Bauman (1964) postulated that the secretion of the subepidermal glands may also function as a sex attractant. Pain and Ruttner (1963) also thought that other pheromones were complementary to queen substance. Butler and Fairey (1964) stated that, "No other substance from these glands (mandibular) we tested was attractive, and we consider that the whole of the olfactory attraction of a queen to drones probably comes from her 9-oxodecenoic and 9-hydroxydecenoic acids."

More evidence is needed to clarify the possibilities that the abdomen is a source of other sex attractants, and that other compounds in the head are functioning as complementary sex attractants. For example, if highly volatile or light-sensitive chemicals were released constantly by living queens, these compounds would not necessarily be measured effectively by the present bioassays, which involve displaying extracts in volatile solvents placed on substrates that facilitate rapid volatilization.

Butler (1967b) reported that 9-oxodec-trans-2-enoic acid also functions as an aphrodisiac in stimulating mounting of the queen by the drone. Yet, Morse and Gary (1962) found that a small percentage of virgin queens without mandibular glands, that were extirpated before they became functional, could nevertheless mate successfully. Even though queen pheromones apparently function as a fundamental means of communication on the mating flight, pheromones from the mandibular glands are not absolutely essential to successful mating.

Inter-specific attraction of drones to chemicals extracted from other queens was demonstrated by Butler et al. (1967). Extracts of A. cerana and A. florea queens were as attractive to A. mellifera drones as extracts of A. mellifera queens.

The queen mating attractant pheromone has been used primarily to facilitate research on drone mating behavior. Queen pheromones displayed in "drone congregation areas" (those relatively rare areas where drones tend to concentrate naturally, or can be lured in great numbers by controlled pheromone release) attract drones to various experimental locations where their activities can be photographed, or they can be captured, labeled, and released for flight range and distribution surveys.

V. Nest Defense

One of the earliest observations pertaining to honey bee pheromones was made by Huber (1814). He discovered that freshly excised stings, or the odor of stings placed near the hive entrance, elicited aggressive attacks by worker bees. Apiculturists have known for many years that the probability of being stung, while manipulating honey bee colonies, increases sharply, and appears to rise exponentially, after the first sting is received. A sweet scent, somewhat similar to banana oil, can be detected at the site of stinging. The sting pheromone was referred to as "alarm" pheromone by Maschwitz (1964) and Wilson (1965).

Ghent and Gary (1962) determined that the source of sting pheromone appeared not to be the venom per se. The secretion seemed to be associated with two masses of glandular cells, lying against the inner surface of the quadrate plates. Individual ducts open onto the outer surface of these plates (Snodgrass, 1956). Release of the secretion is possible by partial extrusion of the sting.

Experimentally, a good bioassay has not been developed to measure the biological activity of the sting pheromone. Attempts by Ghent and Gary (1962) to develop a bioassay under field conditions were abbreviated considerably by a very painful experience from some extremely agitated bees! Under laboratory conditions, bees in a glass-walled observation hive responded quickly to pieces of filter paper, bearing fresh stings, that were inserted into the brood nest. Worker bees became "agitated" near the stings and this response spread quickly outward to a radius of 15-20 cm. Bees were attracted to the paper, and assumed the usual "aggressive" posture that typically precedes aggressive behavior when bees are disturbed at the colony entrance. This behavior has been described by Butler and Free (1952).

Boch et al. (1962) determined that a primary active component in the alarm pheromone is iso-pentyl acetate (iso-amyl acetate). They also found (Boch and Shearer, 1966) that newly emerged worker bees contained essentially no iso-pentyl acetate, while forager bees, 15-30 days old, contained 1-5 µg; caged bees had less than 0.17 µg, and queen bees produced no iso-pentyl acetate. Morse et al. (1967) reported that iso-pentyl acetate is found in all four Apis species.

The alarm pheromone from the sting is apparently supplemented by another chemical, 2-heptanone, found in the worker mandibular glands by Shearer and Boch (1965) and Boch and Shearer (1967). Morse et al. (1967) reported that 2-heptanone is found only in A. mellifera, even though this compound elicits alarm in all members of the genus Apis. Earlier, Maschwitz (1964) had reported that an aggressive reaction by guard bees could be elicited by placing the mandibular gland secretion at the hive entrance. Simpson (1966) suggested that the secretion, when released by guard bees at the nest entrance, may contribute to colony defense by repelling robber bees. There is some evidence to support the hypothesis that 2-heptanone is functional in nest defense; both Simpson (1966) and Butler (1966a) found that foraging honey bees were strongly repelled by the mandibular gland secretion and Butler (1966a) attributed this effect to 2-heptanone. Robber bees are essentially bees foraging on honey stored in other colonies (Gary, 1967), instead of nectar in flowers. Robber bees probably are repelled by 2-heptanone in the same manner as bees that are foraging on other sources. Other insects or small animals conceivably may be repelled by the same pheromone.

Once guard bees are alerted, and stinging is initiated, the primary defense mechanism involves the deposition of stings in the "enemy", severence of the sting structure from the bee's body, and release of the alarm pheromone from the sting. The odor identifies the "victim", elicits further aggressive behavior, and facilitates orientation to the "enemy".

Future research on possible alarm pheromones in the

venom should be facilitated greatly using refined methods
developed by Gunnison (1966) for collecting large quantities
(mls) of bee venom. Research on colony defense pheromones
is severely handicapped by the absence of an effective bioassay. An assay is needed in which moving objects, presented in an appropriate visual context, elicit flight and
stinging behavior by guard bees. A short flight is necessary inasmuch as pheromone-impregnated objects placed
in contact with the nest entrance may elicit some elements
of house cleaning behavior that could be confused with
aggressive behavior associated with nest defense. Furthermore, the objects should have an optimum physical consistency that not only permits easy penetration of the sting, but
also facilitates retention of the barbs. Otherwise, guard
bees, regardless of the state of alarm or aggregation, cannot deposit stings.

    To my knowledge, the use of alarm pheromones in prac-
tical bee management has not been considered seriously.
Theoretically, it should be possible to induce adaptation
and/or habituation in bees by releasing alarm pheromones
inside colonies. By this means the aggressive stinging
behavior of bees possibly could be minimized or eliminated
during colony manipulations, provided a technique is
developed to control the initial aggressive response when
the pheromone is released inside the colony.

VI. Pheromones Associated with Foraging

    Pheromones associated with the foraging activities in
the field seem to originate primarily, if not solely, from
the Nassanoff gland (scent gland), located dorsally on
segment VI of the worker abdomen. The secretion from several hundred unicellular glands accumulates in a pouch-like,
invaginated membrane (Jacobs, 1925) which, when exposed,
permits rapid evaporation of the secretion. Some foraging
worker bees release the Nassanoff gland pheromone (and
perhaps other pheromones) near food sources, and this odor
is extremely attractive to other foragers. Von Frisch
(1923) demonstrated this behavior in an experiment in
which experienced foraging bees, with their Nassanoff
glands sealed by shellac painted over the tergites, attracted approximately 10% as many new recruits as normal bees.
Also, von Frisch (1923) found that pheromone release from
the scent gland was stimulated greatly by the more rewarding

food sources, where copious sugar syrup was available to
foragers; less rewarding food stations did not stimulate
Nassanoff gland eversion, and consequently attracted approximately 10% as many new recruits, when compared with the
more rewarding food stations.

Apparently the frequency of exposure of the Nassanoff
gland increases as the odor concentration associated with
the food decreases. Adrian M. Wenner (personal communication) observed frequent gland exposure of bees feeding on
unscented sucrose syrup, and significantly less gland
exposure when clove oil was added to the syrup.

Renner (1960) collected the Nassanoff gland secretion
by wiping exposed scent glands with filter paper. He
determined that the secretion is highly attractive, but is
neither colony nor race specific. The first active compound that was identified in the Nassanoff gland secretion
was geraniol (Boch and Shearer, 1962). In addition, they
discovered (Boch and Shearer, 1963) that worker bees
started producing geraniol when they were approximately
two weeks old; bees of foraging age contained approximately
1.0 µg per bee. Later, Free (1962) determined that the
whole secretion was considerably more attractive than
geraniol alone, thus indicating the probable presence of
other bioactive components in the pheromone. Later, Boch
and Shearer (1964) identified nerolic and geranic acids as
additional active components, and reported that the three
compounds appeared to account for most of the attraction.
However, Weaver et al. (1964) identified still another
active component, citral (geranial), and concluded that a
mixture of geraniol and citral is apparently as attractive
as the whole Nassanoff secretion.

After further investigation, Shearer and Boch (1966)
reported that citral was absent in freshly collected
Nassanoff gland secretion, but that citral developed when
the whole secretion was held at room temperature. Later,
Butler and Calam (1969) analyzed Nassanoff gland secretion
within 4 minutes of its collection and found both isomers
of citral present to the combined extent of approximately
3% of the amount of geraniol present. The research to date
indicates that citral is the primary attractant, but that
its attractiveness is enhanced by some or all of the other
constituents in the secretion, particularly geraniol. In
summary, research on the Nassanoff gland secretion seems
to be a good example to emphasize the hazards involved in

data interpretation, and the severe problems encountered when using bioassay testing procedures to evaluate fractions, and combinations of fractions, that normally interact in a complex pheromonal secretion.

It is difficult to interpret the foregoing experiments in terms of their significance in developing an understanding of function of the Nassanoff gland pheromone, as related to normal foraging behavior; most of the observations were made at feeding stations that do not necessarily simulate natural flora, in terms of the quality and quantity of nectar, or its spatial or temporal distribution in flowers. Adrian M. Wenner (personal communication) has challenged the notion that Nassanoff gland pheromones are used by worker bees that are foraging on normal food sources; he postulates that pheromone release behavior may be caused by the special conditions present at feeding stations. He further postulates that the primary function of the Nassanoff gland pheromone may be the release of alighting behavior by flying bees.

Although the frequency of Nassanoff gland exposure by foraging bees in natural conditions is apparently very low, the Nassanoff gland pheromones nevertheless may be quite functional in marking unusually rewarding food sources, especially in broad areas as opposed to individual plants. Perhaps the Nassanoff gland pheromone functions as an intra-species and inter-colony communication that confers a great advantage to the species, rather than to individual colonies. This hypothesis is predicated on the suppositions that (a) during evolution of the bee, available nest sites were sparse and constituted a primary limiting factor in regulating the population density of honey bees; and that (b) inter-specific competition for food was far more significant than inter-colony competition. Therefore, the fact that the Nassanoff gland pheromone is not colony-specific is not sufficient evidence to minimize its possible role in scent-marking of food sources.

There is some speculation concerning the possible applications of synthetic Nassanoff pheromone in training bees to visit marginally attractive crops to effect pollination. Although the possibilities for application are relatively unexplored, there is an abundance of empirical evidence indicating that the primary stimuli regulating foraging activity are associated with the quality and quantity of available nectar and/or pollen. Odors from plants or bees are significant in attracting bees to a food source, and enhancing the overall stimulus value of the food. Nevertheless, sustained foraging activity and active recruitment of new foragers is motivated by consistent reward, i.e., the consistent availability of rich food sources.

The use of queen pheromones to stimulate foraging activity is a promising application. Elbert R. Jaycox (personal communication) has found interesting preliminary evidence that indicates a relationship between nectar foraging activity of colonies and the presence of queen pheromones. Small test colonies with queen pheromones supplied as synthetic 9-oxodecenoic acid, whole queen extracts, or living queens, show greater foraging activity than queenless colonies. Furthermore, the mortality of worker bees is greater under the stress of queen pheromone deprivation. Showers (1967) observed that the introduction of synthetic queen substance into small, queenless colonies, apparently retained a higher percentage of workers than similar colonies without queen substance.

Expendable, queenless colonies are being considered seriously as a substitute for normal colonies when bees are needed in locations where pesticide exposure is too severe to risk loss or damage to normal colonies.

VII. Swarming and Swarm Migration

Swarming and supersedure behavior apparently are related to the production and distribution of queen pheromones. Butler (1960b) found that superseded and "swarm" queens from meagerly-populated colonies had approximately 25% as much queen substance as mated, laying queens. In densely-populated colonies, apparently preparing to swarm, he determined that queen pheromone production was apparently normal; inefficient collection and/or distribution of queen pheromones among workers must have caused the loss of the queen cell-building inhibition.

Although swarms may leave the colony, move a short distance, and cluster temporarily, the pheromone complex of the queen is essential for normal clustering and swarm migration. A swarm that leaves the colony without the queen soon returns to the colony. Under certain conditions, colonies can be induced to swarm and return daily. Morse (1963) demonstrated that swarms "followed" a caged queen

held in his hand. Morse (1963) and Simpson (1963) also
showed that pheromones from the head of the queen caused a
cohesion of the swarm. Simpson (1963) determined that
synthetic queen substance (9-oxodec-trans-2-enoic acid)
does not affect swarm clustering behavior. Butler et al.
(1964) found, under controlled environmental conditions,
that 9-hydroxydecenoic acid stimulated swarms to form and
maintain a stable cluster.

In subsequent experiments by Butler and Simpson (1967),
the complementary nature of 9-oxodec-trans-2-enoic acid
and 9-hydroxydec-trans-2-enoic acid was discovered, that is,
the former was strongly attractive to flying bees from a
queenless swarm while the latter stimulated bees to alight
and group into a "quiet" cluster. A mixture of these acids
was as attractive as the pheromones from living, mated
queens.

The Nassanoff gland pheromone is used in orientation
communication, especially in swarms. When swarming bees
enter a new nest site they characteristically elevate their
abdomens, expose their Nassanoff glands, and fan vigorously
as they slowly proceed in a chain-like "procession" into
the nest opening. Presumably the basic mechanism is orien-
tating into the direction of air flow, produced by anterior
bees, that are fanning and releasing pheromone. The same
response can be observed when bees from a swarm are placed
on a smooth substrate, such as paper, and a caged queen is
placed upwind. Queen pheromones also release the fanning
and scenting behavior, and the bees become aligned in the
general direction of the queen, even from a distance of
several feet from the queen cage. Within a few minutes the
"procession" of bees forms a compact cluster around the
caged queen and fanning soon ceases.

There is a possibility that honey bee pheromones
could be used to prevent swarming. Swarm prevention is a
primary objective in modern bee management. By treating
colonies with synthetic queen pheromones during the early
phases of "preparations" for swarming, it may be possible
to block certain behavior, such as queen cell construction,
that is requisite for swarming.

VIII. General Considerations of Further Research
on Honey Bee Pheromones

Chemical analyses of queen bees, and parts of queens,
reveal a large number of compounds that may function as

pheromones (Callow et al., 1964; Pain et al., 1962). Also, there is a substantial amount of empirical behavioral evidence, in many publications not cited in this review, that indicates the probable existence of other unknown pheromones in honey bees. This seems especially likely in regards to intra-colony behavior where the population density and great variety of behavioral interactions create intensive selective pressure on the evolution of increasingly efficient communications systems, paralleling the evolution of more and more sophisticated social organization. In honey bees, efficient distribution of pheromones is provided by food-sharing activities, frequent physical contacts, and regulated air circulation.

The most significant barrier to further research on honey bee pheromones is, as usual, the development of sensitive and valid bioassays to measure the appropriate biological activities. One major problem is that bee bioassays should be performed under normal environmental conditions. However, this requirement presents almost insurmountable problems when intra-nest bioassays are conducted in social insects, especially in honey bees. The "normal" social environment presents an abundance of uncontrolled variables, such as (a) heterogeniety of age, physiological condition, and genetic composition; (b) the presence of a multitude of unknown pheromones, related directly and indirectly to the experimental pheromone and present in unknown amounts; (c) an enclosed nest in which light, necessary for observations, may change the critical behavioral criteria; and (d) the unique difficulties in utilizing miniaturized automatic data recording instrumentation in a nest environment where bees destroy instruments by chewing or interfere with their operation by coating surfaces with propolis and beeswax. When pheromone complexes are being tested, especially when the relative amounts of each of the constituent compounds determine the quality and quantity of information, there is the additional risk of uncontrolled changes in ratios of the chemicals during testing procedures. For example, highly volatile chemicals displayed on a substrate are lost more quickly than low vapor pressure compounds, thus altering the ratios of compounds in a complex mixture. Partial solutions to some of these problems of testing pheromones are: (a) if possible, keep all materials frozen until immediately before bioassay; (b) utilize living specimens that provide a constant source of

pheromones; (c) use freshly excised materials, such as whole
glands which can be ruptured at the moment of testing; and
(d) minimize the test duration as much as possible.

IX. Controlling Behavior of Bees and
Other Insects by Pheromonal Manipulation

It seems likely that the most promising potential
control of insect behavior by pheromone manipulation may be
found in those species that have evolved the greatest dependence upon pheromonal communication, as opposed to
other stimulus modes. However, this does not necessarily
imply that the species utilizing larger numbers of pheromones, especially in complex combinations as in the honey
bee, will be more susceptible to behavioral regulation by
pheromonal manipulation. Neither should one always expect
an insect species utilizing a simple pheromone (single
compound) which elicits a single behavioral act, to be mani-
pulatable by synthetic pheromones. Obviously, the degree
of man-imposed control over the various species will depend
primarily upon (a) the number of available alternate pathways, that is, the utilizable stimulus modes in the total
stimulus configuration that stimulate the behavior; (b)
the innate behavioral plasticity, or learning capabilities;
and (c) the genetic adaptability, or ability of the insect
to alter its behavior and pheromonal systems by genetic
mechanisms.

The natural resistance of insects to man-modified
pheromonal elements in the environment probably will parallel the resistance problem that has been experienced with
pesticides. Concomitantly, we may expect that man-manipu-
lated pheromones will be extremely useful and valuable in
insect control, and perhaps as significant as many insecticides that are currently used. Control of insects by
pheromonal manipulation offers an additional advantage,
however, over pesticide control. As insects develop different pheromonal pathways that facilitate essential biological objectives, the new pheromonal key to the control
problem is present in the new population, and can be
sought directly. Yet, it is conceivable that many insect
species may be selected for the capacity to utilize completely non-pheromonal mechanisms of communication, even
though the present populations of these species may utilize
pheromonal systems extensively.

CONTROL OF INSECT BEHAVIOR

As man struggles to regulate insect activities by manipulating pheromones, the following considerations should be kept in mind: (a) multiple-compound pheromone complexes seem to be biological insurance and probably have more communicative potential. Therefore, such pheromone complexes should be anticipated as a frequent occurrence in insects, especially when there are complex, prolonged, behavioral interactions under conditions of high population density; (b) threshold perception levels are often phenomenally low (for example, queen substance), suggesting that insects have evolved "selective sensitivity" to compounds produced and utilized within the species, and that less volatile compounds may be more significant than previously supposed; (c) physiological responses should be given more attention in bioassay procedures, to complement the usual behavioral responses; (d) investigators should consider the total stimulus complex or configuration in the test animals' environment, rather than attempting to regulate an entire behavioral pattern by manipulating a single variable in the environment. For example, learning phenomena, such as habituation, make it appear unlikely that the massive release of a single pure compound will permanently alter the behavior of an entire population; (e) researchers should strive to resist the temptation to state that, "I have discovered the pheromone, mechanism, or stimulus that is responsible for eliciting an observed behavior". The pheromonal key that seems to fit the biological lock usually has many other stimuli coded into its notches and grooves.

X. References

Barbier, M., and Hügel, M. F. (1961). Bull. Soc. Chim. Fr. 202, 951.

Barbier, M., and Lederer, E. (1960). C. R. Acad. Sci. 250, 4467.

Barbier, M., and Pain, J. (1960). C. R. Acad. Sci. 250, 3740.

Barbier, M., Lederer, E., and Nomura, T. (1960). C. R. Acad. Sci. 251, 1133.

Boch, R., and Shearer, D. A. (1962). Nature 194, 704.

Boch, R., and Shearer, D. A. (1963). J. Insect Physiol. 9, 431.

Boch, R., and Shearer, D. A. (1964). Nature 202, 320.

Boch, R., and Shearer, D. A. (1966). J. Apicult. Res. 5, 65.

Boch, R., and Shearer, D. A. (1967). Z. Vergleich. Physiol. 54, 1.

Boch, R., Shearer, D. A., and Stone, B. C. (1962). Nature 195, 1018.

Butler, C. G. (1954a). "The World of the Honeybee". Collins, London.

Butler, C. G. (1954b). Trans. R. Ent. Soc. Lond. 105, 11.

Butler, C. G. (1956). Proc. R. Ent. Soc. Lond. (A) 31, 12.

Butler, C. G. (1960a). Experientia 16, 424.

Butler, C. G. (1960b). Proc. R. Ent. Soc. Lond. (A) 35, 129.

Butler, C. G. (1961). J. Insect Physiol. 7, 258.

Butler, C. G. (1966a). Nature 212, 540.

Butler, C. G. (1966b). Z. Bienenforsch. 8, 143.

Butler, C. G. (1967a). Biol. Rev., Cambridge 42, 42.

Butler, C. G. (1967b). Proc. R. Ent. Soc. Lond. (A) 42, 71.

Butler, C. G., and Calam, D. H. (1969). J. Insect Physiol. 15, 237.

Butler, C. G., and Fairey, E. M. (1963). J. Apicult. Res. 2, 14.

Butler, C. G., and Fairey, E. M. (1964). J. Apicult. Res. 3, 65.

Butler, C. G., and Free, J. B. (1952). Behaviour 4, 263.

Butler, C. G., and Gibbons, D. A. (1958). J. Insect Physiol. 2, 61.

Butler, C. G., and Paton, P. N. (1962). Proc. R. Ent. Soc. Lond. (A) 37, 114.

Butler, C. G., and Simpson, J. (1958). Proc. R. Ent. Soc. Lond. (A) 33, 120.

Butler, C. G., and Simpson, J. (1965). Scientific Studies, Univ. Libcice, Czechoslovakia 4, 33.
Butler, C. G., and Simpson, J. (1967). Proc. R. Ent. Soc. Lond. (A) 42, 149.
Butler, C. G., Calam, D. H., and Callow, R. K. (1967). Nature 213, 423.
Butler, C. G., Callow, R. K., and Chapman, J. R. (1964). Nature 201, 733.
Butler, C. G., Callow, R. K., and Johnston, N. C. (1959). Nature 184, 1871.
Butler, C. G., Callow, R. K., and Johnston, N. C. (1961). Proc. R. Soc. (B) 155, 417.
Callow, R. K., and Johnston, N. C. (1960). Bee World 41, 152.
Callow, R. K., Chapman, J. R., and Paton, P. H. (1964). J. Apicult. Res. 3, 77.
Erp, A. van (1960). Insectes Sociaux 7, 207.
Free, J. B. (1962). J. Apicult. Res. 1, 52.
Frisch, K. von (1923). Zool. Jb. (Physiol.) 40, 1.
Gary, N. E. (1961a). Ann. Entomol. Soc. Am. 54, 529.
Gary, N. E. (1961b). Science 133, 1479.
Gary, N. E. (1961c). Bee World 42, 14.
Gary, N. E. (1962). Science 136, 773.
Gary, N. E. (1967). Amer. Bee J. 106, 446.
Gary, N. E., and Morse, R. A. (1962). Proc. R. Ent. Soc. Lond. (A) 37, 76.
Gast, R. (1967). Insectes Sociaux 14, 1.
Ghent, R., and Gary, N. E. (1962). Psyche 69, 1.
Groot, A. P. de, and Voogd, S. (1954). Experientia 10, 384.
Gunnison, A. F. (1966). Proc. R. Ent. Soc. Lond. (A) 37, 76.
Huber, F. (1814). "Nouvelles Observations sur les Abeilles II". Transl. 1926. Hamilton, Ill. Dadant.
Jacobs, W. (1925). Z. Morphol. M. Okol. 3, 1.
Johnston, N. C., Law, J. H., and Weaver, N. (1965). Biochemistry 4, 1615.
Maschwitz, V. W. (1964). Nature 204, 324.
Morse, R. A. (1963). Science 141, 357.
Morse, R. A., and Gary, N. E. (1962). Nature 194, 605.
Morse, R. A., and Gary, N. E. (1963). Ann. Entomol. Soc. Am. 56, 372.
Morse, R. A., Shearer, D. A., Boch, R., and Benton, A. W. (1967). J. Apicult. Res. 6, 113.
Pain, J. (1961a). C. R. Acad. Sci. 252, 2316.
Pain, J. (1961b). Annls. Abeille 4, 73.

Pain, J., and Barbier, M. (1960). C. R. Hebd. Séanc. Acad.
   Sci., Paris 250, 1126.
Pain, J., and Ruttner, F. (1963). C. R. Hebd. Séanc. Acad.
   Sci., Paris 256, 512.
Pain, J., Barbier, M., Bogdanovsky, D., and Lederer, E.
   (1962). Comp. Biochem. Physiol. 6, 233.
Pain, J., Barbier, M., and Roger, B. (1967). Annls. Abeille
   10, 45.
Renner, M. (1960). Z. Vergleich. Physiol. 43, 411.
Renner, M., and Bauman, M. (1964). Naturwissenschaften
   51, 68.
Shearer, D. A., and Boch, R. (1965). Nature 206, 530.
Shearer, D. A., and Boch, R. (1966). J. Insect Physiol. 12,
   1513.
Showers, R. E. (1967). Amer. Bee J. 107, 294.
Simpson, J. (1963). Nature 199, 94.
Simpson, J. (1966). Nature 209, 531.
Snodgrass, R. E. (1956). "Anatomy of the Honey Bee".
   Cornell Univ. Press, Ithaca, New York.
Velthuis, H. H. W., and Es, J. van (1964). J. Apicult.
   Res. 3, 11.
Velthuis, H. H. W., Verheijen, F. J., and Gottenbos, A. J.
   (1965). Nature 207, 1314.
Verheijen-Voogd, C. (1959). Z. Vergleich. Physiol. 41, 527.
Voogd, S. (1955). Experientia 11, 181.
Weaver, N., Weaver, E. C., and Law, J. H. (1964). Prog. Rep.
   Tex. Agric. Exp. Sta. No. 2324, 1.
Wilson, E. O. (1965). Science 149, 1064.
Zmarlicki, C., and Morse, R. A. (1964). Ann. Entomol. Soc.
   Am. 57, 73.

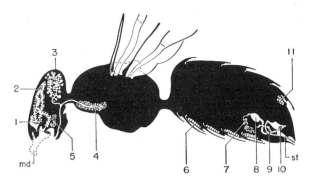

Figure 1.--Exocrine gland systems of the honey bee worker
that are sources, or potential sources, of pheromones.
(1) Mandibular gland; (2) hypopharyngeal gland; (3)
head labial gland; (4) thorax labial gland; (5) post-
genal gland; (6) wax glands; (7) poison gland; (8)
vesicle of poison gland; (9) Dufour's gland; (10)
Koschevnikov's gland; (11) Nassanoff's gland. (Adapted from Wilson, 1965).

Figure 2.--"Retinue" behavior of worker bees attracted to
the pheromones of the queen bee.

51

Figure 3.--Single queen bees confined in individual cages
 are bioassayed to determine the relative attraction of
 worker bees to the queens. A lid is placed on top of
 the hive during the test to provide darkness and to
 eliminate extraneous air currents. The relative
 attraction of queens is assessed by the numbers of
 workers that cling to the cages, after removal from
 the hive.

Figure 4.--A cluster of drones, attracted to queen pheromones, is hovering momentarily below and on the lee-
ward side of a cage containing a living virgin queen.
There are 34 drones (some are partially blocked from
view). Note that the drone nearest to the cage (left
side) is being mounted by another drone.

THE SUBSTANCES THAT CONTROL THE FEEDING BEHAVIOR AND
THE GROWTH OF THE SILKWORM Bombyx mori L.

Yasuji Hamamura

Biological Department
Kōnan University
Motoyama, Kōbe Japan

Table of Contents

Table of Contents--(Continued)

CONTROL OF INSECT BEHAVIOR

I. Introduction

The silkworm larva, Bombyx mori L., has been regarded as one of the most typical examples of a monophagous insect. Although it can be experimentally fed on some other Moraceae and other plants, it feeds almost exclusively on mulberry leaves. In this paper the author intends to discuss several substances responsible for the feeding behavior and growth of the silkworm larva, and finally synthetic diets without mulberry leaves. Studies were begun by the author in 1935 (Hamamura, 1959).

II. Feeding Behavior

At the beginning of this work, volatile substances contained in mulberry leaves were considered to stimulate the appetite of the silkworm. Fresh leaves were extracted with cold ethanol, and the attractancy of the extract was tested as follows: a filter paper impregnated with the extract was placed on one side of a box (7 x 35 cm) and the amorphous, brown residue on the opposite side. Silkworm larvae (fifth instar) were released in the center of the box, and the box was covered. After 30 min, all larvae were attracted to the filter paper, but none of them tried to bite it. Thus, the cold-ethanol-soluble substances were responsible only for the attraction of larvae. The residue, however, was eaten by larvae when they were placed on it. Agar threads soaked in a hot methanol extract of the residue were vigorously bitten by larvae. Although an agar jelly mixed with this extract was not continuously eaten by the larvae, continuous feeding occurred when the agar jelly containing the methanol extract was mixed with the residue.

From these results, it was concluded that there are 3 important factors responsible for the feeding behavior of the silkworm larva: an attractant, a biting factor, and a swallowing factor.

III. Fractionation from Leaves

In later experiments it was found that the methanol extract contained 2 biting factors, one ether soluble and one water soluble. The residue still showed biting activity in addition to a swallowing factor. The method for separation of these factors was thus modified as shown on the next page:

57

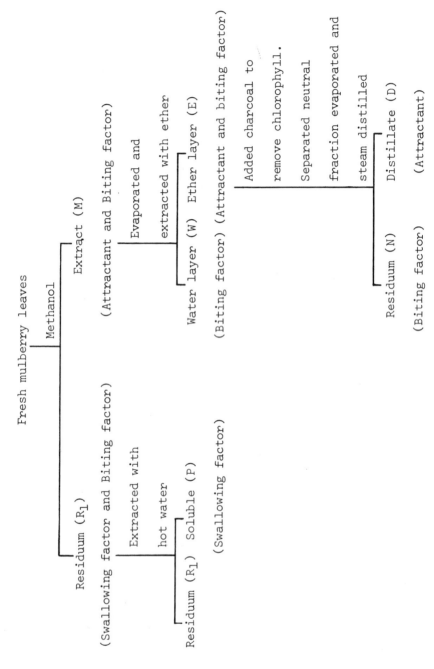

Fresh mulberry leaves

Methanol

Residuum (R₁)
(Swallowing factor and Biting factor)

Extract (M)
(Attractant and Biting factor)

Extracted with hot water

Residuum (R₁)
(Swallowing factor)

Soluble (P)
(Swallowing factor)

Evaporated and extracted with ether

Water layer (W)
(Biting factor)

Ether layer (E)
(Attractant and biting factor)

Added charcoal to remove chlorophyll. Separated neutral fraction evaporated and steam distilled

Residuum (N)
(Biting factor)

Distillate (D)
(Attractant)

CONTROL OF INSECT BEHAVIOR

IV. Attractants

A. Biological Test

The following biological test was used for evaluation
of the activity. Six small circles were drawn along the
circumference of a round filter paper (8.3 cm dia) at equal
intervals, and the paper was placed in a petri dish. Wet
cellulose powder on which the sample was absorbed, and controls (wet cellulose only) were placed alternately in the
small circles. Thirty newly hatched larvae were placed on
the center of the filter paper. After 10, 30 and 60 min,
the larvae on each portion and those wandering on the intermediate space were counted. The tests were carried out at
25°C.

B. Isolation and Characterization of Active Principles

The methanol extract (M) was dissolved in ether after
evaporation of methanol, and activity was found in the ether
layer (E). This fraction was treated with active carbon to
remove chlorophyll, and was washed successively with dilute
sulfuric acid and a dilute sodium carbonate solution. The
neutral fraction separated was steam distilled.
Strong activity was found in the steam distillate
(Table I). A chromatogram of the distillate suggested the

Table I

Attractiveness of the Steam Distillate to B. mori Larvae

Area of filter paper	Numbers of larvae attracted after		
	10 min	30 min	60 min
Steam distillate	13	23	25
Residue	7	3	1
Control	5	1	0
Intermediate space	5	3	4
Total	30	30	30

59

presence of citral, linalyl acetate, linalool and terpinyl acetate. The activities of these compounds found in mulberry leaves were tested using authentic samples. These terpenoids were found to be very attractive to larvae, but other terpenoids which have not been found in mulberry leaves were not active in the concentrations shown in Table II.

Table II

Attractiveness of Several Terpenoids[a] to B. mori Larvae

Terpenoid	Numbers of larvae attracted after 1 hr at 24°C			
	Sample	Control	Intermediate	Total
Terpinyl acetate[b]	13	6	11	30
Lynalyl acetate[b]	17	12	1	30
Linalool[b]	20	0	10	30
Citral[b]	13(20)	13(7)	4(3)	30
Terpineol	2	15	13	30
Geraniol	0	12	18	30
Geranyl acetate	3	15	12	30
Hexenol	13	4	13	30

[a] Concentration of sample about 30 µg, except 6 µg in parentheses.
[b] Contained in mulberry leaves.

Torii (1948) found that one palpus responded to an attractant stimulus, and another to a feeding stimulus. Only hexenol has been found as an attractant for larvae since our study (Watanabe, 1958). Ishikawa (1965) recently confirmed our results using an olfactometer, and discovered many other attractants for the larvae.

CONTROL OF INSECT BEHAVIOR

V. Biting Factors

A. Biological Test

The agar thread, on which the sample is absorbed, is useful for testing biting activity because it could not be bitten off. Thus, the biting action could easily be observed. Counting fecal pellets did not distinguish between biting and swallowing activities.

B. Isolation and Identification

The fractions (M), (W), (E), (R_1) and (N) were found to be active (Table III). Apparently several different substances are involved in the biting activity.

A crystalline substance was obtained by Hamamura et al. (1961) from the ether soluble fraction (N) containing one of the biting factors. This substance was found to be identical with β-sitosterol which had already been isolated from mulberry leaves by Naito and Hamamura (1961a). Later, Goto et al. (1965) found β-sitosterol β-glucoside and lupeol in mulberry leaves. These substances, especially β-sitosterol, play an important role in the biting response (Table IV). β-Sitosterol is contained in the waxy substances covering the upper surface of mulberry leaves.

The water soluble biting factor in the fraction (W) (Hamamura et al., 1962a) was identified as isoquercitrin, which had already been found in mulberry leaves by Hayashiya and Hamamura (1956). The biting activity initiated by several flavones is shown in Table V. Isoquercitrin induces less biting activity than does β-sitosterol.

Other flavonoids, such as rutin, quercetin and morin, were also tested, but only morin showed activity. Morin was found in the stem of Morus tinctoria, but not in mulberry leaves. Inositol and sucrose were isolated from the water soluble fraction (W) by Naito and Hamamura (1961b), but neither substance elicited biting activity. But these 2 compounds are indispensable for the feeding of silkworm larvae, as discussed later. The effect of β-sitosterol on the nutrition (Ito, 1960a,b) and feeding behavior (Ito, 1961) of silkworm larvae has been reported earlier. B. mori (Fukuda et al., 1962) and Phillosamia cynthia ricini Iones (Fukuda, 1964) have been reared on artificial diets containing β-sitosterol.

61

Table III

Biting Response of B. mori Larvae to Fractions
from Fresh Mulberry Leaves

Fraction	Biting Response[a,b]
M	+ + + + +
W	+ + + + -
E	+ + + - -
E + W	+ + + + +
R_1	+ + + + +
R_2	+ + - - -
D	- - - - -
N	+ + + + -
P	- - - - -

[a] + + + + + All larvae immediately showed biting response.
   + + + + - Almost all larvae showed biting response.
   + + + - - Larvae showed weak biting response.
   + + - - - Only 1 or 2 showed biting activity.
   - - - - - No larvae responded.
[b] Ten fifth instar larvae were used in each test.

CONTROL OF INSECT BEHAVIOR

Table IV

Biting Response of <u>B</u>. <u>mori</u> Larvae to Several Sterols

Sterol[a]	Biting Response[b]
β-Sitosterol	+ + + + +
β-Sitosterol glucoside	+ + + + -
Cholesterol	+ + + - -
Lupeol	+ + + - -
Ergosterol	+ + - - -

[a] The agar threads were sprayed with an ether solution of the sterols until the surfaces of the agar threads were covered with sterol crystals. Then water was sprayed on the threads to soften them.

[b] + + + + + All larvae immediately showed biting response.
+ + + + - Almost all larvae showed biting response.
+ + + - - Larvae showed weak biting response.
+ + - - - Only 1 or 2 showed biting activity.

Table V

Biting Response of B. mori Larvae to Several Flavones
and the Other Substances Found in Fraction W

Test Substance[a]	Biting Response[b]
Isoquercitrin	+ + + - -
Morin	+ + + + -
Rutin	- - - - -
Quercetin	- - - - -
Sucrose	- - - - -
Inositol	- - - - -

[a] The agar threads were soaked in 0.5% solution of flavones, or 3% solution of sugars.
[b] + + + + - Almost all larvae showed biting response.
+ + + - - Larvae showed weak biting response.
- - - - - No larvae responded.

Nayar and Fraenkel (1962) denied the effect of β-sitosterol on the feeding behavior of silkworm larvae, but Hamamura (1963) could not accept their opinion. Horie (1962) also recognized the activity of isoquercitrin and morin in the feeding of silkworm larvae.

Hayashiya (1966) reported that the 2 hydroxyl groups located in carbon-2' and -3 of the flavone nucleus (I) are indispensable for biting activity.

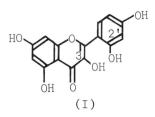

(I)

VI. Swallowing Factors

A. Biological Test

The swallowing activity was evaluated by counting the number of faeces excreted. The basal diet (Bd) was as follows: 50 ml of 3% agar solution mixed with 5 g of sugar for taste, and 5 g cellulose powder for bulk. Samples to be tested were added to this basal diet, and 5 fifth instar larvae were placed on the diet.

B. Isolation and Identification

Diets Nos. 5 and 6 (Table VIII), which consisted of "Bd + M + R_1" and "Bd + M + P", respectively, seemed to be sufficient for feeding activity of larvae. ((M) is the methanol soluble substances of mulberry leaves containing the attractant and biting factors, (R_1) the methanol insoluble substances containing the swallowing factors, and (P) the hot water extract from (R_1).) As seen in Tables VI and VIII, (M) can be partially but not completely replaced by β-sitosterol.

Fraction (M) can be replaced by a mixture of β-sitosterol and another biting factor, isoquercitrin or morin, in addition to sugar and inositol. These compounds were all found in fraction (M), and inositol and sugar were confirmed as swallowing factors (Table VII).

65

Table VI

Influence of Sterols on Swallowing and
Biting Activities of B. mori

Diet[a]	Number of faeces[b]
1 Basal diet (Bd) only[c]	26
2 Bd + M + R_1	262
3 Bd + β-Sitosterol + R_1	223
4 Bd + β-Sitosterol + Cellulose	41
5 Bd + Cholesterol + R_1	16
6 Bd + Ergosterol + R_1	44

[a] Dosages: R_1, 0.3 g; each sterol, 20 mg.
[b] Number of faeces excreted from 10 newly hatched larvae
was counted after 18 hr at 24°C.
[c] Basal diet consisted of 5 ml of 2% agar solution containing 1% sucrose and 0.2 g cellulose.

Table VII

Feeding Activity of B. mori on Diets Containing Several Substances Found in Fraction M^a

Diet	Number of faeces
Basal diet (Bd) only	0
Bd + M + R_1	192
Bd + β-Sitosterol + R_1	89
Bd + β-Sitosterol + Inositol + R_1	200
Bd + β-Sitosterol + Morin + R_1	206
Bd + β-Sitosterol + Inositol + Morin + R_1	254
Bd + β-Sitosterol + Inositol + Isoquercitrin + R_1	226
(Bd minus Sucrose)+ β-Sitosterol + Inositol + Morin + R_1	60

[a] M contained the biting factor and attractant; however, the attractant was not necessary as the larvae were placed directly on the diet.

If larvae were placed directly on the diet, the larvae fed on the diet as shown in Table VII. In this case, attractants in fraction (M) were useless for orientation of larvae. The principal component of (R_1) was thought to be cellulose. If cellulose powder was removed from Bd of No. 6, the number of faeces decreased from 186 to 30 (Table VIII).

From these results, cellulose seemed to be one of the main swallowing factors. But (R_1) could not be replaced completely with chemically pure cellulose powder as already shown in No. 4 of Table VI. In addition to cellulose powder, the boiling water extract (P) or a glacial acetic acid extract from (R_1) was necessary for the feeding of larvae.

Hamamura et al. (1962b) detected large amounts of silica and potassium phosphate in the boiling water extract (P). Thus, it was found that cellulose coated with silica gel and potassium phosphate was effective for eliciting the swallowing response. But (R_1) could not yet completely be replaced with these substances (Table IX). Mukaiyama and Ito (1962) confirmed the importance of cellulose for the feeding response of the silkworm.

From these experimental results, the substances controlling the feeding behavior of silkworm larvae were almost clarified. The agar jelly containing these substances in the proportions shown in Table X, was eaten by the larvae almost as avidly as were mulberry leaves themselves.

It is concluded that silkworm larvae are attracted by compounds such as citral contained in the diet, and bite because of the presence of β-sitosterol, which is further enhanced by morin. Sugar, inositol, cellulose, silica, and phosphate promote continuous feeding.

C. Discussion

These substances concerned in the feeding of silkworm larvae are not found exclusively in mulberry leaves, but are rather common generally in green leaves. The preference for mulberry leaves may depend on the amounts and proportions of these compounds, and on the absence of repellents. To test for repellents, small amounts of raw soybean cake, powdered milk or chlorella, etc., were added to artificial diets. Larvae were indeed repelled. Extraction of these additives with methanol removed the repellent components.

Table VIII

Feeding Activity of $\underline{B.\ mori}$ on Diets Containing Various
Fractions With and Without the Swallowing Factor

Exp. No.	Diet[a]	Number of faeces[b]
1	Basal diet (Bd) + Leaf powder (5g)	204
2	Fresh leaves	199
3	Bd + M	78
4	Bd + E	36
5	Bd + M + R_1	190
6	Bd + M + P	186
7	Bd + W + E + R_1	192
8	Bd + P	40
9	Bd + R_1	21
10	Bd only	24
11	(Bd minus Cellulose) + M + P	30

[a] Dosage of each fraction corresponded to 5 g of powdered mulberry leaves.
[b] Faeces were counted after rearing overnight at 25°C.

Table IX

Feeding Activity of B. mori on Chemically Pure Diets

Diet	Number of faeces
Basal diet (Bd) only	0
Bd $+$ M $+$ R_1	192
Bd $+$ β-Sitosterol $+$ Inositol $+$ Morin $+$ Phosphate $+$ R_1	204
Bd $+$ β-Sitosterol $+$ Inositol $+$ Morin $+$ Phosphate	127
Bd $+$ β-Sitosterol $+$ Inositol $+$ Morin $+$ Phosphate $+$ Silica	162

Table X

Synthetic Diet for B. mori

Component	Quantity
Citral[a]	1 ml (10 mg% ether solution)
β-Sitosterol	5 mg
Isoquercitrin or morin	3 mg
Cellulose powder	700 mg
Sucrose	30 mg
Inositol	5 mg
Potassium phosphate	10 mg
Silica gel	40 mg (coated in cellulose)
2% Agar solution	3 ml

[a] Citral or other attractants were not necessary when larvae were placed directly on the diet.

VII. Growth Factors

Fukuda et al. (1962) and Matsubara et al. (1967)
succeeded in rearing silkworm larvae using several artificial diets containing mulberry leaf powder. But if mulberry
leaf powder was not added to the diet, the growth of silkworm larvae was much retarded and only a few larvae produced
poor cocoons.

Our studies focused particularly on the substances in
mulberry leaves that promoted larval growth of the silkworm. The water soluble fraction (W) and ether soluble
fraction (E) were found to be important not only for feeding
behavior but also for the growth of larvae.

Naito and Hayashiya (1965) isolated chlorogenic acid
(II) from fraction (W), and Kato and Yamada (1963-64) also
found chlorogenic acid as a water soluble growth factor.
Yamada and Kato (1966a) confirmed these experimental results.

(II)

Recently Yamada and Kato (1968) reported the following
isomers of chlorogenic acid in mulberry leaves: isochloro-
genic acid, neochlorogenic acid and another acid having a
band at 510 mμ in U.V. These isomers were found to be
growth factors as well as chlorogenic acid itself.

Hamamura et al. (1966) found that gallic acid showed
growth promoting activity on early larval stages. Okauchi
et al. (1968) found recently that protocatechuic acid and
many other polyphenolic acids and aldehydes showed the
same effect as gallic acid.

When 0.5% of gallic acid was added to diet A (Table
XI), the larval weight after 15 days from hatching was
156 mg, whereas the control was 95 mg. The weight of the
cocoon produced was about 400 mg, as large as that produced
on a normal diet. In mulberry leaf-free diets, larval
growth and cocoon weight were poor compared with the diet A.

Table XI

Composition of Synthetic Diets for B. mori

Substance	Diet A	Diet B
Mulberry leaf powder	20 g	–
Soybean protein	25 g	20 g
Starch	15 g	15 g
Sucrose	10 g	10 g
Cellulose powder	35 g	50 g
β-Sitosterol	500 mg	500 mg
Vitamin mixture	500 mg	400 mg
Wesson's mineral	1000 mg	900 mg
Ascorbic acid	500 mg	500 mg
Inositol	500 mg	500 mg
K_2HPO_4	500 mg	500 mg
Choline chloride	50 mg	50 mg
Distilled water	160 ml	150 ml

Kato and Yamada (1966a) demonstrated that 1-substituted 3,4-dihydroxybenzene structure (III) in the polyphenolic acid was essential for promoting growth of larvae.

(III)

Yamada and Kato (1966a, 1967) isolated linolenic and palmitic acids from the ether soluble fraction and found that the addition of both acids to the diet promoted larval growth. Oleic acid had the same effect as both acids, although this acid was not found in leaves (Table XII).

Ito and Nakasone (1966) and Ito et al. (1966) also reported on the nutritive value of some oils and fatty acids for the silkworm. Kato and Yamada (1966b) found that several kinds of oils such as cotton seed and peanut oil showed noticeable effects on growth promotion, and chlorogenic acid promoted the utilization of these oils. The basal diet used in these biological tests is shown in Table XIII.

Hayashiya et al. (1963a,b) found that royal jelly was a useful substance in rearing larvae on a diet free from mulberry leaves, and later, Hayashiya et al. (1965) found that the active principle in royal jelly was acetylcholine. Kato and Ishiguro (1964) demonstrated that acetylcholine stimulated molting of larvae.

VIII. Synthetic Diet

From our findings on feeding stimuli and growth factors, Yamada and Kato (1966b) proposed an improved diet, free from mulberry leaves, including chlorogenic acid and oleic acid as shown in Table XIII.

Kato and Yamada (1967) confirmed that body weight, growth rate and cocoon production of larvae were improved by addition of chlorogenic acid and oleic acid to the diet (Table XIV).

Thus, silkworm larvae can be reared without mulberry leaves but the weight of the cocoon layer is still not satisfactory. We are now trying to find the additional components responsible for growth promotion and production of cocoons.

Table XII

Comparative Growth-Promoting Activity of Oleic Acid and a
Combination of Linolenic and Palmitic Acids

Diet additive	Survival (per cent)	Third instar after 10 days (per cent)	Weight (mg)
None	67	47	27.8
1% Oleic acid	92	85	36.0
2% Oleic acid	96	96	83.1
Linolenic + Palmitic acid	100	87	48.6
1% fatty acid mixture[a]	100	92	66.1

[a] Fatty acid fraction from mulberry leaves.

Table XIII

Composition of the Basal Diet Used by
Yamada and Kato (1966b)

Substance	Quantity
Cellulose powder	5.2 g
Soy bean protein	2.0 g
Potato starch	1.5 g
Sucrose	1.0 g
Wesson's salt mixture	0.09 g
Vitamin mixture[a]	0.04 g
Inositol	0.05 g
K_2HPO_4	0.05 g
β-Sitosterol	0.05 g
Choline chloride	0.05 g
Oleic acid	0.02 g
Chlorogenic acid	0.05 g
Protocatechuic acid	0.05 g
Water	15.0 ml

[a] Hayashiya et al. (1963a).

Table XIV

Rearing of Silkworm Larvae on a Synthetic Diet Containing
Oleic and Chlorogenic Acids[a]

	Diet with oleic and chlorogenic acids	Diet with fresh mulberry leaves
Larvae matured	89%	93%
Larval period	29 da	24 da
Yield of cocoons	78%	93%
Cocoon mean weight	1.89 g	1.80 g
Cocoon shell mean weight	377.4 mg	409.9 mg
Cocoon shell ratio	20.0%	22.8%
Yield of pupae	78%	87%

[a] From Kato and Yamada (1967).

77

IX. Summary

The feeding behavior of the silkworm larva depends
sequentially on 3 factors: attractant, biting and swallowing. The larva initially responds to the attractant contained in mulberry leaves, then bites on stimulation by the
biting factor, and finally feeds on stimulation by the
swallowing factor.
    The following compounds were found in mulberry larves:
Attractants: citral, linalyl acetate, linalool and terpinyl
    acetate.
Biting factors: β-sitosterol, isoquercitrin and morin.
Swallowing factors: cellulose, sugar, inositol, silica and
    potassium phosphate.
    The larva feeds on an agar jelly containing all these
compounds, and, with the addition of several nutrients, is
able to develop. However, growth and development are poor
unless 20% of the diet is mulberry leaf powder. Several
growth promoting factors were found in the mulberry leaves,
and these were identified as chlorogenic acid, linolenic
acid and oleic acid. Moreover, acetylcholine was found to
stimulate the molting of larvae. Polyphenolic acids, such
as gallic acid or protocatechuic acid, also have a growth
promoting effect on the young stages of larvae.
    Thus, silkworm larvae can be reared on the synthetic
diet free of mulberry leaves, but the weight of the cocoon
layer is still not satisfactory.

X. References

Fukuda, T. (1964). Bull. Sericult. Exp. St. 19, 201.
Fukuda, T., Sudo, M., Kameyama, T., and Kawasugi, S. (1962).
J. Agr. Chem. Soc. Japan 36, 819.
Goto, M., Imai, S., Murata, T., Fujioka, S., Fujita, E.,
and Hamamura, Y. (1965). Ann. Rep. Takeda Res. Lab.
24, 55.
Hamamura, Y. (1959). Nature 183, 1746.
Hamamura, Y. (1963). Kagaku to Seibutsu (Japan) 1, 364.
Hamamura, Y., Hayashiya, K., and Naito, K. (1961). Nature
190, 881.
Hamamura, Y., Kuwata, K., and Masuda, H. (1966). Nature
212, 1386.
Hamamura, Y., Hayashiya, K., Naito, K., Matsuura, K., and
Nishida, J. (1962a). Nature 194, 754.
Hamamura, Y., Hayashiya, K., Naito, K., Matsuura, K., and
Nishida, J. (1962b). Nature 194, 756.
Hayashiya, K. (1966). Bochu-Kagaku (Japan) 31, 137.
Hayashiya, K., and Hamamura, Y. (1956). J. Agr. Chem. Soc.
Japan 30, 361.
Hayashiya, K., Kato, M., and Hamamura, Y. (1965). Nature
205, 620.
Hayashiya, K., Naito, K., Nishida, J., and Hamamura, Y.
(1963b). J. Agr. Chem. Soc. Japan 37, 735.
Hayashiya, K., Naito, K., Matsuura, K., Nishida, J., and
Hamamura, Y. (1963a). J. Agr. Chem. Soc. Japan 37, 160.
Horie, Y. (1962). J. Sericult. Sci. Japan 31, 258.
Ishikawa, S. (1965). Bull. Sericult. Exp. St. 20, 21.
Ito, T. (1960a). J. Insect Physiol. 5, 95.
Ito, T. (1960b). Proc. Japan Acad. 36, 287.
Ito, T. (1961). Nature 191, 882.
Ito, T., and Nakasone, S. (1966). Bull. Sericult. Exp. St.
(Tokyo) 20, 375.
Ito, T., Horie, Y., Arai, N., Watanabe, K., Nakasone, S.,
and Shinohara, E. (1966). Sansi-Kenkyu 56, 27.
Kato, M., and Ishiguro, T. (1964). Proc. Japan Acad. 40,
131.
Kato, M., and Yamada, H. (1963-64). Silkworm 10-11, 85.
Kato, M., and Yamada, H. (1966a). Proc. Japan Acad. 42,
1185.
Kato, M., and Yamada, H. (1966b). Life Sci. 5, 717.
Kato, M., and Yamada, H. (1967). Proc. Japan Acad. 43, 234.
Matsubara, F., Kato, M., Hayashiya, K., Kodama, R., and
Hamamura, Y. (1967). J. Sericult. Sci. Japan 36, 37.

Mukaiyama, F., and Ito, T. (1962). J. Sericult. Sci. Japan 31, 398.

Nayar, J. K., and Fraenkel, G. (1962). J. Insect Physiol. 8, 505.

Naito, K., and Hamamura, Y. (1961a). J. Agr. Chem. Soc. Japan 35, 848.

Naito, K., and Hamamura, Y. (1961b). J. Agr. Chem Soc. Japan 35, 1106.

Naito, K., and Hayashiya, K. (1965). J. Agr. Chem. Soc. Japan 39, 237.

Okauchi, T., Kamada, M., and Hamamura, Y. (1968). Ann. Meeting Appl. Entomol. Zool. (Tokyo, 4 April).

Torii, K. (1948). Bull. Sericult. Inst. 2, 3.

Watanabe, T. (1958). Nature 182, 325.

Yamada, H., and Kato, M. (1966a). Proc. Japan Acad. 42, 399.

Yamada, H., and Kato, M. (1966b). Proc. Japan Acad. 42, 1189.

Yamada, H., and Kato, M. (1967). Proc. Japan Acad. 43, 230.

Yamada, H., and Kato, M. (1968). 38th Nat. Meeting of Sericult. (Nagoya).

SENSORY RESPONSES OF PHYTOPHAGUS LEPIDOPTERA
TO CHEMICAL AND TACTILE STIMULI

Frank E. Hanson

Department of Zoology
University of Texas
Austin, Texas

Table of Contents

I. Introduction

One of the main themes of this symposium is that insects recognize certain objects (food, mates, etc.) by detecting and responding to specific chemical stimuli. This is well illustrated by phytophagous insects, particularly the oligophagous ones whose diet is restricted to one or a very few families of plants. Usually this insect-host dependence is obligatory: in the absence of the host the insect starves to death. Thus, it is not surprising that chemical components of the host plants serve as strong attractants and phagostimulants to the dependent insects, whereas those of non-host plants are often repellent.

In order to unravel the complexities of the insect-host plant relationship, a good understanding of the basic physiological mechanisms of these strong phagostimulant/phagodeterrent systems is necessary. With such knowledge, the potentialities of controlling oligophagous insect populations by the manipulation of these stimuli can be more accurately assessed.

The purpose of this paper is to summarize current physiological knowledge bearing on chemosensory mechanisms of phytophagous lepidopteran larvae. Also included will be recent findings from our laboratory which concern the tactile receptors on the mouthparts of these insects and the extent to which mechanosensory information may play a role in feeding. Special attention will be directed toward insects with a narrow host range since these extreme cases may be more instructive in determining the physiological principles involved in feeding than would those insects with a broad host range.

II. Responses Of Gustatory Chemoreceptors To Host Plants

Physiological data from phytophagous insects stimulated by plant substances have become available only in the last decade, and are consequently somewhat sketchy. However, enough data exist to indicate that no single physiological mechanism has been adopted to maintain specific feeding behavior in these insects. The following discussion concerns three such mechanisms.

A. Phagostimulation by Secondary Plant Substances

Some insects depend on secondary plant substances for

82

attraction and stimulation. These chemicals apparently have
no physiological importance to the plant, and therefore are
tentatively called "secondary". They are highly diverse
and appear in a great variety of concentrations. For
example, von Rudloff et al. (1967) point out that coastal
and inland species of juniper (Juniperus ashei Buchholz
and Juniperus virginiana L.) have 80-90 different terpenes,
only ten of which are common to these allopatric species.
With such a rich variety of plant chemicals available, it
is not surprising that they are used by specific feeders
as phagostimulant cues. One worker (Fraenkel, 1959) even
suggested that the raison d'etre of these substances is
plant-insect interaction.

As early as 1910 Verschaffelt noted that in order for
feeding to occur in some species of Pieris, mustard oil
glucosides, which are minor constituents of their host
plants, were required. He further suggested that "these insects are to a great extent guided in their choice of plant-
food by the presence of such glucosides". The Pieris case
has been further documented by David and Gardiner (1966a,b),
who reported that P. brassicae (L.) fed more frequently on an
artificial diet which contained mustard oil glucosides than
on one which did not. The threshold concentration at which
stimulation first occurred was not determined, but feeding
is definitely increased at 3×10^{-7} M sinigrin and is further
enhanced up to the highest concentration tested, 3×10^{-4} M.
All of the nine glucosides tested stimulated feeding. Most
of the test glucosides have been identified in the natural
host (Brassica) or in an alternate host (Tropaeolum).

The physiological basis of this feeding stimulation was
elucidated by Schoonhoven (1967). Responses of chemosensory cells located on the mouthparts were elicited by
solutions containing mustard oil glucosides at 10^{-5} M and
10^{-6} M. Two such cells per maxilla were discovered, one in
each of the sensillum styloconicum. Each cell is responsive
to different (but overlapping) sets of mustard oil glucosides. The specificity of these receptors for glucosides
is not absolute: sodium cyanate is also a stimulant,
although it is required at concentrations 100 times higher.

B. Phagostimulation by Nutritive Compounds

Whereas secondary plant substances confer no benefit
to the insect other than some behavioral cue, some plant
components are nutritive as well as phagostimulative.

For example, sucrose stimulates feeding in P. brassicae (Schoonhoven, 1967). The effect was shown to be additive to the phagostimulation produced by the mustard oil glucosides. A correlated electrophysiological study demonstrated that a chemoreceptive cell in the maxillary sense organs responds to sucrose. This receptor is found in only the lateral sensillum styloconicum and is distinct from the cell responsive to mustard oil glucosides. Thus summation in the CNS of inputs from two distinct receptors, which respond to different compounds, results in an increase in the probability of a feeding response.

Another example of an insect with a narrow host range, which may be stimulated to feed by the nutritive compounds in the leaf, is the silkworm, Bombyx mori L. The electrophysiological data of Ishikawa (1963) and Ishikawa and Hirao (1963) showed that, out of a total of eight receptor cells, three responded to sugars, one to salts and acids, and one to water. The sugar receptors were quite sensitive: the threshold of the "sucrose receptor" was 10^{-4} M sucrose, while that of the "inositol receptor" was 1.5×10^{-4} M inositol. Various sugars stimulated the former, but the latter was specific to inositol. A third cell typically responded to D-glucose at about 2×10^{-3} M. All of these are contained in the mulberry leaf, which is the natural food of B. mori. When a mixture of these sugars at the same concentrations in which they occur in the leaf is applied to the receptors, the impulse patterns are similar to those resulting from stimulation by aqueous extracts of the mulberry leaf itself. Thus a considerable amount of the chemosensory information obtained by the insect from the mulberry leaf concerns nutrients.

The behavioral studies of Y. Hamamura reported in this symposium confirm that sugars do stimulate feeding, although they probably act in concert with other stimulants.

C. Phagodeterrence by Repellent Substances

The amount of feeding in insects may be regulated mainly by the ratio of phagodeterrents to phagostimulants. This was demonstrated in B. mori, by the behavioral studies of Ishikawa and Hirao (1966). Their earlier electrophysiological studies (Ishikawa and Hirao, 1963) showed that only one of eight chemosensory cells responded dependably to aqueous extracts of leaves from several plant species.

Accordingly, these authors concluded that this cell responds
only to repellent substances, and thus governs the insects'
rejection of all leaves except those of mulberry. The study
did not exclude the possibility that other receptor cells
may be sensitive to repellents as well. Indeed, it would be
remarkable if only one cell is responsive to the repellent
substances of all leaves other than mulberry.

III. Responses Of Tactile Receptors
On The Mouthparts To Host Plants

The foregoing discussion has presumed that chemoreceptors are of primary importance in feeding. Tactile
receptors may also play a role in feeding, such as monitoring the consistency of the leaf or portions thereof. Such
a mechanism could cause rejection of leaf veins which has
been observed for many phytophagous insects.

The mouthparts of caterpillars are amply supplied by
tactile receptors. The larva of the tobacco hornworm,
Manduca sexta (Johan.), has five prominent setae on the
maxilla (Fig. 1), as described by Schoonhoven and Dethier
(1966). A study of the innervation of the area by the
same authors indicated that other areas, such as the eight
campaniform sensilla, the sensilla stylonconica, and "free"
nerve endings between segments of the maxillary palpus,
may be mechanoreceptive as well. Preliminary results of an
electrophysiological study (Hanson, unpublished data) confirm
the existence of mechanoreceptor activity in the suggested
areas. Figure 1 is a diagrammatic representation of the
maxilla showing the various electrode placements and directions of movements during the application of stimuli.

A. Tactile Setae

These large hairs on the maxillae (Fig. 1) appear
well formed for use as tactile probes, and electrophysiological studies indicate that these setae function as
mechanoreceptors. To demonstrate this, a sharpened tungsten electrode was inserted via a micromanipulator down
the center of the stump of a cut seta (Fig. 1). Movements
of the electrode in turn moved the seta, producing the
response shown in Fig. 2. The axis of displacement most
sensitive to movement was the vertical one; forces causing

both compression and elongation produced phasic and tonic responses. Translational and radial movements produced no responses except when obvious compression or stretching also occurred. These data illustrate the mechanoreceptive characteristics of the setae, and confirm the results reported by Ishikawa and Hirao (1963) for B. mori.

B. Sensilla Styloconica

At the tip of each sensillum styloconicum is a small papilla (Fig. 1) which can be easily deflected without apparent movement of the sensillum. Even minor deflections produce a response, which lasts as long as pressure is applied, in one of the cells in the sensillum (Fig. 3). The three widely separated spikes are the response to 0.1 M NaCl in a glass pipette positioned over the sensillum. The recording in Fig. 3 was obtained from a tungsten needle inserted into the base of a sensillum styloconicum (Fig. 1 at B). On the basis of the size and shape of the resulting neural impulse, the responding cell does not appear to be one of the chemoreceptive cells; however, this is difficult to prove. Thus when leaf sap is applied by means of small pipettes, a small burst of this unique impulse can often be seen while the somewhat tacky sub-tance is withdrawn from the sensillum (Fig. 4). Occasion-ally it is also present at the beginning of a stimulus, but usually the intense activity of the chemoreceptors obscures this part of the record.
    These data indicate the presence of a tonic mechano-receptor whose sensitive membrane is present in this ter-minal papilla. The histological work of Schoonhoven and Dethier (1966) shows a group of cell bodies below the sensillum styloconicum and dendritic extensions running up inside the sensillum to the papilla. One of these is apparently sensitive to mechanical deformations. Whether it is also a chemoreceptor remains to be investigated.

C. Maxillary Palpus

Movement of the segments of the palpus relative to one another activates mechanoreceptors located near the joints. To study this, electrosharpened tungsten needles were inserted through the soft tissue between the segments on the lateral side of the maxillary palpus (Fig. 1). Bending, pushing, or pulling produces responses in the receptor even though the movement is only a few microns (Figs. 5 and 6). A tonic response from the sustained application of pressure is seen in Fig. 6. No activity is recorded when the joints assume natural positions.

Thus it appears that the mechanoreceptors of the maxillary palpi are capable of monitoring the degree to which the palpi telescope in response to compressive forces. By correlating this information with the force of contraction bringing the two contralateral maxillae together while palpating a leaf edge or pieces of leaf bitten off by the mandibles, the insect can undoubtedly determine some information about the consistency of the potential food.

When one considers the relative paucity of neurons and other sense cells in this insect, it appears that the mouthparts of M. sexta larvae are fairly well endowed with mechanoreceptors. The animal undoubtedly uses these tactile receptors for the coordination of mouthparts and the manipulation of food while feeding. It is probable that these systems can be used to determine the consistency of the potential food, and so may function as another physiological mechanism which maintains feeding specificity.

Acknowledgments

The experiments reporting electrophysiological data on mechanoreceptors of M. sexta were performed in the laboratory of Dr. V. G. Dethier then of the Department of Zoology, University of Pennsylvania. The author gratefully acknowledges the assistance and facilities generously offered by Dr. Dethier, as well as the assistance of Dr. Tibor Jermy of the Research Institute for Plant Protection, Budapest, Hungary.

IV. References

David, W. A. L. and Gardiner, B. O. C. (1966a). Entomol. Exp. Appl. 9, 95.
David, W. A. L. and Gardiner, B. O. C. (1966b). Entomol. Exp. Appl. 9, 247.
Fraenkel, G. S. (1959). Science 129, 1466.
Ishikawa, S. (1963). J. Cell. Comp. Physiol. 61, 99.
Ishikawa, S. and Hirao, T. (1963). Bulletin of the Sericultural Experiment Station (Tokyo) 18, 336.
Ishikawa, S. and Hirao, T. (1966). Bulletin of the Sericultural Experiment Station (Tokyo) 20, 320.
Schoonhoven, L. M. (1967). Koninkl. Nederl. Akademie van Wetenschappen - Amsterdam. Proceedings, Series C, 70, 556.
Schoonhoven, L. M. and Dethier, V. G. (1966). Archives Neederlandaises de Zoologie 16, 497.
Verschaffelt, E. (1910). Proc. Acad. Sci. Amsterdam 13, 536.
von Rudloff, E., Irving, R. and Turner, B. L. (1967). Am. J. Bot. 54, 660.

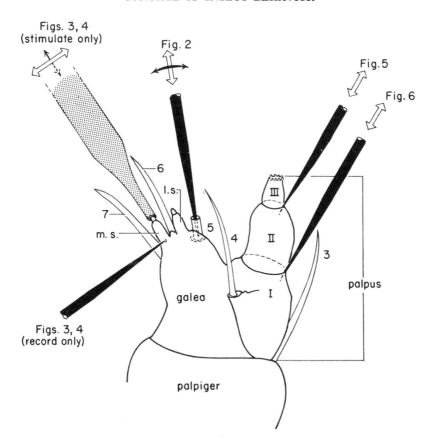

Figure 1.--The maxilla of Manduca sexta (after Schoonhoven
    and Dethier, 1966). m.s., medial sensillum stylo-
    conicum; l.s., lateral sensillum styloconicum; 3-7,
    tactile setae (5 has been cut to allow the insertion
    of an electrode); I-III, segments of the maxillary
    palpus. Electrode placements are shown with the
    corresponding figure of the response record. The axes
    of movement which elicit responses are indicated by
    arrows, the larger arrow being the more effective
    directions of movement. Broken arrow (B) indicates
    probable stimulation.

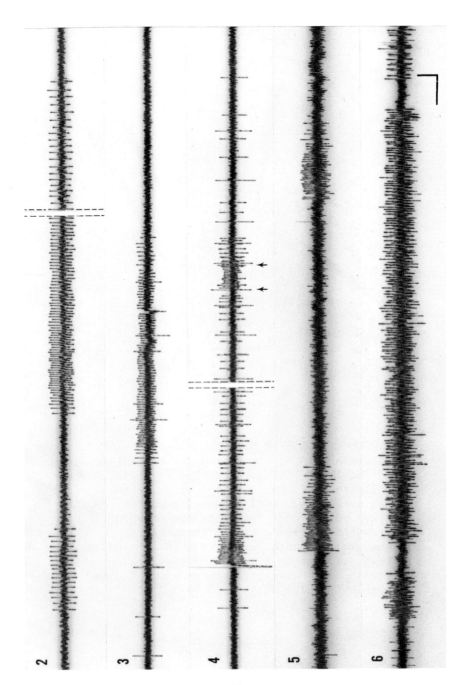

Figures 2-6.--Electrophysiological records of responses of
maxillary mechanoreceptors to movement. In all cases,
responses last only for the duration of the stimulus.
Time calibration (horizontal bar): 0.1 seconds; voltage
calibration (vertical bar): 0.5 mv. for figures 2-4,
0.2 mv. for figures 5-6.

Figure 2.--Response of large seta to vertical movements;
0.42 seconds of the tonic response deleted (dashed
lines).

Figure 3.--Response of papilla on tip of sensillum styloconicum; stimulated chemically by 0.1 M NaCl (few
large spikes) in glass pipette, and mechanically by
moving pipette laterally against the papilla.

Figure 4.--Same as Figure 3, but stimulated chemically by
exudate of cabbage (Brassica), and mechanically
(arrows) by the withdrawal of electrode containing
this tacky' fluid. Between arrows is the burst of
small spikes of the mechanoreceptor neuron which are
unlike any of the chemoreceptor spikes seen in the
record; 0.2 seconds of the record deleted (dashed
lines).

Figure 5.--Response of palpus segmental mechanoreceptor.
Recording electrode is between palpus segments II and
III. Stimuli are brief compression or extension for-
ces operative at the segmental junction.

Figure 6.--Response of palpus segmental mechanoreceptor.
Recording electrode is between palpus segments I and
II. A phasic response to a brief compression stimulus and a tonic response to a stimulus of longer
duration are seen.

AGGREGATION OF THE GERMAN COCKROACH,
Blattella germanica (L.)

Shoziro Ishii

Pesticide Research Institute
College of Agriculture
Kyoto University
Kyoto, Japan

Table of Contents

Table of Contents--(Continued)

CONTROL OF INSECT BEHAVIOR

I. Introduction

Gregariousness is common in many species of insects.
It has also been found in domiciliary cockroaches such as the
American cockroach, Periplaneta americana (L.) and the
German cockroach, Blattella germanica (L.) (Wille, 1920;
Gould and Deay, 1938; Ishii and Kuwabara, 1967). Gregariousness varies in intensity according to species, and within
a species with the age or physiological state of the insects.
Aggregation favors the growth and development of cockroach nymphs (Landowski, 1938; Willis et al., 1958; Wharton
et al., 1967). It is believed that aggregation depends on
the olfactory response to chemical substances produced by
cockroaches themselves (Pettit, 1940; Ledoux, 1945). Roth
and Willis (1960) comprehensively reviewed the subject of
gregariousness in cockroaches and emphasized that cleverly
designed laboratory experiments will be required to elucidate the nature of this behavior.
In the course of our study we found that gregariousness
of the nymphs depends largely on certain chemical sub-
stance(s) contained in their faeces and body surface, and
that this substance serves as an attractant to the cockroach for aggregation.

II. Growth Pattern of Nymphs

Growth and development of German cockroach nymphs were
studied under both solitary and gregarious conditions.
Approximately 40 newly hatched nymphs from one oötheca were
divided into the following 2 groups:
Solitary: 20 newly hatched nymphs were reared individually
on dog biscuit (Oriental Yeast Co. Ltd.) and water in small
petri dishes, 2.8 cm dia x 1.2 cm high.
Gregarious: 20 newly hatched nymphs were reared together on
the same medium in a petri dish, 9 cm dia x 2 cm high. The
food was renewed every 2-6 days to avoid contaimination with
molds. Feeding experiments were carried out at a temperature of 25° ± 2°C under 14 hr illumination alternating with
10 hr darkness daily. Each experiment was replicated with
2 oöthecae.
The period from egg hatch to maturation of nymphs which
were reared in groups was shorter than that for those reared
individually (Table I). Gregariousness is more noticeable
during the first and second instars than in later instars.

Table I

Nymphal Period from Egg Hatch to Adult Emergence of
B. germanica Reared Individually and in Groups of 20 Nymphs

		Numbers of adults emerged ♂♂ ♀♀		Mean nymphal period (days)	Mean body weight (mg) ♂♂ ♀♀	
A	Solitary	11	(6 5)	65.6 ± 2.4	49	67
	Grouped	17	(6 11)	58.3 ± 2.0	52	71
B	Solitary	14	(7 7)	65.7 ± 2.6	51	71
	Grouped	14	(7 7)	55.6 ± 2.2	50	74

Also ecdyses tended to be simultaneous in the grouped individuals. There was no significant difference in the body
weight of adults reared under solitary or gregarious conditions.

In order to determine the effect of density on the
growth and development of nymphs, newly hatched nymphs were
reared in solitary and in groups of 2, 5, and 10 nymphs.
Rearing conditions were almost the same as in the previous
experiment. Each experiment was replicated 3 times.

The effect of density on growth and development was
observed even when nymphs were reared in groups of only 2
individuals, and there were no significant differences in
the length of the nymphal period for groups of 2, 5, and 10
nymphs (Table II). There did not appear to be any differences in body weight between those reared individually and
those reared in groups.

III. Mode of Aggregation

When a piece of filter paper which had been used for a
shelter in rearing a group of cockroaches for several days
was put into a glass pot in which nymphs had been released,
the nymphs tended to aggregate on this filter paper rather
than on clean filter paper. A 3-choice experiment to

Table II

Development of B. germanica Nymphs
Reared Individually and in Groups

Numbers of nymphs in group	Numbers of adults emerged	Nymphal period (days)	Mean body weight (mg) ♂♂	♀♀
1	10(1)[a]	63.5±3.2	47.9±2.1	63.0
	9(4)	66.1±4.2	56.6±5.9	76.3±8.7
	10(6)	66.2±2.5	56.3±4.7	73.3±5.8
2	8(4)	58.0±2.8	43.3±2.4	67.0±5.9
	8(4)	57.8±3.1	53.3±8.0	76.3±7.7
	8(6)	57.4±2.1	52.0	75.7±1.4
5	9(3)	58.1±4.5	45.0±2.9	62.7±2.1
	10(6)	57.7±2.4	55.3±1.7	72.7±5.9
	8(3)	54.5±1.9	48.6±4.0	77.0±4.9
10	10(6)	55.6±2.6	46.8±5.2	65.0±4.0
	9(6)	58.6±1.8	56.7±2.8	75.7±3.1
	9(6)	59.6±7.7	50.3±7.7	76.0±3.2

[a] Figures in parentheses indicate numbers of females.

assess the aggregation of nymphs was carried out using conditioned filter paper, 3.5 cm x 7.5 cm, folded in a W-shape, and 2 clean filter papers of the same size. The results are shown in Figs. 1-4. This aggregation was also observed in darkness. When the antennae of the nymphs were cut off, no aggregation was observed. These findings indicate that aggregation was induced by a response to chemical stimuli. The conditioned filter paper was contaminated with faeces and perhaps other odors emitted from the cockroaches. When faeces collected from another batch of German cockroaches were affixed with CMC (carboxymethyl cellulose) to a filter paper, the nymphs aggregated on this paper in the same way as on the naturally conditioned paper. A filter paper impregnated with ether or methanol extract of faeces had the same effect. However, faeces excreted from silkworm larvae, Bombyx mori (L.), had no such effect. Thus, it seems that a certain substance(s) responsible for aggregation is contained in their faeces, and is perceived through their antennae.

IV. Localization of Active Principle

In preliminary experiments, a filter paper impregnated with concentrated ether washings of the body surface of German cockroaches was found to elicit the aggregation response. Since some secretions were found on the body surface of cockroaches, it was necessary to determine whether or not the active principle found in faeces originates in the secretion from the body surface.

A. Biological Assay

The following 2-choice method was adopted for the biological assay. A small piece of filter paper, 3.5 cm x 7.5 cm, folded in a W-shape, was conditioned with faeces or the extract to be tested. Clean filter paper of the same size was used as a control. These 2 filter papers were placed in opposite positions in a glass pot, 11 cm dia x 7 cm high. Some 20-30 first instar nymphs were introduced into the center of the pot. The biological assay was carried out under the conditions previously described. The assay was usually started in the evening and the nymphs were allowed to move about freely until next morning without food or water. Each test was replicated 10 times.

B. Site of Active Principle.

In order to localize the active principle, 10-20 adult males were cut into 5 portions, head, legs, wings, thorax and abdomen. Each portion was washed with ether, and 10 pieces of filter paper, impregnated with the ether washings, were assayed.

The results indicated that the active principle is mainly localized in the abdomen (Table III). To determine whether the active principle might have been extracted from internal abdominal tissues through the cut opening of the abdomen, the abdomen was ligatured between its first and second segments with a fine silk thread prior to cutting. The ether wash of the ligatured abdomen elicited the same response as that from a non-ligatured one.

Table III

Aggregation Response to Filter Papers Impregnated with
Ether Washings of Several Portions of
Adult Male B. germanica

| | Response[a,b] | | |
	+	±	-
Head	22	14	14
Legs	20	14	16
Wings	28	11	11
Thorax	11	12	27
Abdomen	40	7	3

[a] 5 replications.
[b] + most nymphs aggregated on the conditioned paper;
± about half of the nymphs aggregated on the conditioned paper and half on the control, or, almost none of them aggregated on either the conditioned or the control paper;
- most of nymphs aggregated on the control.

To determine the site of production in the abdomen, 20 adult male abdomens were transversely cut between the third and sixth abdominal segments (mostly through the fifth). These 2 portions were washed 4 times each with a small amount of ether and the washings were pooled. Ten pieces of filter paper were impregnated with the washings from the anterior and 10 with the washings from the posterior portions, and assayed. Ether washings from the anterior and posterior portion were compared simultaneously. The test was replicated twice.

The ether washings from the posterior portions were more active than washings from the anterior portions. As the posterior abdomen seemed to be the site of production, 20 posterior portions of abdomens were further transversely cut into 2 portions, and each portion was washed with ether and assayed as before. When the posterior portion of the abdomen was cut between the seventh and ninth segments, the ether washing of the terminal segments (7-10) was found to be the most active.

These data showed that the active principle is concentrated on the surface of the terminal segments of the abdomen as well as in their faeces.

C. Separation of Faeces after Excretion

It was important to decide whether the active principle was primarily produced externally in the region of the anus and then absorbed onto the faeces, or whether it was primarily contained in the faeces and then absorbed onto the body surface.

Since the cockroaches used for the previous experiments were reared in a box, they were continuously contaminated with their own faeces. In order to prevent this, the following experiment was performed. Ten adult males were reared at 25° ± 2°C for 1 week on dog biscuit and water in a small cage, 5 x 5 x 10 cm, made from 16-mesh wire screen. The cage was suspended from the top of an insect rearing case, 27 x 25 x 42 cm, and a petri dish was placed on the bottom of the case to collect the faeces. One adult excreted about 10 faecal pellets during 24 hr under these conditions.

The faeces were extracted with ether and the extract assayed. The caged adults were washed with ether, and this washing was also tested. Both the extracts of faeces and the body washings were found to be attractive for the nymphs. These studies suggested that the active principle(s) was produced in the region of the anus, either externally or internally.

V. Site of Secretion

The abdomen of the adult male consists of 10 segments, as shown in Fig. 5. A pair of depressions is located on the eighth abdominal tergite, where secretion occurs at courtship (Wille, 1920; Roth and Willis, 1960). These depressions are found only in the adult male and not in the female or the nymphs. The secretion from the tergal glands lying beneath the epidermis of the depression cannot function in aggregation, because faeces excreted from females and nymphs

have the same effect as faeces from males. As secretory
organs responsible for aggregation could not be found externally in the region of the anus of the male cockroach by
microscopic observation, histological studies were undertaken.

A. Histological Study

The rectum and reproductive organ are the 2 principal
organs in abdominal segments 8-10. The fact that faeces
excreted from both sexes of the adult and from nymphs are
equally effective, suggests that the reproductive organs
and their accessories are not the sites of secretion. Thus,
attention was focused on the hind-gut and the anus.
Bouin's solution was mainly used for fixation of the
organs, and paraffin blocks were made in the usual manner.
Transverse and longitudinal sections, 5 to 14µ thick, were
made, and the sections were stained with haematoxylin-eosin,
haematoxylin-phloxin, and paraldehyde-fuchsin solutions.
It was found that the epithelium of the anterior part
of the rectum formed 6 rectal pads consisting of a single
layer of cells covered with a thin layer of the intima. The
cells comprising the rectal pads are columnar in shape and
elevate on the side toward the lumen. They have large
nuclei and appear to be secretory in nature. No such secretory cells were found in the posterior part of the rectum.
Rectal pads were found in both sexes and also in the nymph.
The structure of the cells in the nymph was the same as that
of the adult except that they were smaller in number and
size. The rectum is about 1.6 mm in length and 0.7 mm in
width in the male, and somewhat larger in the female. Anal
glands found in some insects could not be found in the
German cockroach. The posterior part of the colon, just
anterior to the rectum, is a narrow canal, with epithelial
cells which are different in structure from those in the
rectal pads, suggesting no secretory function.

B. Role of the Rectum

If faeces could be excreted without passing through the
rectum, it might be possible to separate the role of the
rectum and the surface of the cuticle in the production of
the substance causing aggregation. A fine glass tube was
inserted from the anus into the intestine to permit faeces

to bypass the rectum. However, faeces were not excreted in this experiment.

A transverse cut was made between the eighth and the ninth abdominal segments of males to remove their abdominal tips where the rectum is located. These cockroaches were released in a glass pot where 10 small pieces of filter paper had been placed. Conditioning of filter papers with amputated cockroaches was carried out by supplying dog biscuit and water for 13-48 hr. After 2 days almost all the insects had died. In a subsequent assay, first instar nymphs did not exhibit an aggregation response to the conditioned filter papers (Table IV).

Table IV

Aggregation Response to Filter Papers Conditioned with Adult Male B. germanica whose Abdominal Tips had been Amputated

Numbers of insects for conditioning	Time for conditioning (hr)	Response[a] +	±	-
20	13	0	3	7
40	18	1	3	6
50	24	0	1	9
50	48	1	6	3

[a] + most nymphs aggregated on the conditioned paper;
± about half of the nymphs aggregated on the conditioned paper and half on the control, or, almost none of them aggregated on either the conditioned or the control paper;
- most of nymphs aggregated on the control.

The rectum was dissected from the amputated abdominal tips. They were then ground with a small glass rod and extracted with a small portion of methanol. Filter papers were immersed into the extract and after evaporation of the solvent the papers were assayed. The nymphs aggregated mostly on the papers impregnated with the methanol extract of the rectums (Table V).

Table V

Aggregation Response of B. germanica to Filter Papers Impregnated with Methanol Extract of the Rectum

Numbers of rectums	Numbers of rectums containing faecal matter	Response[a]		
		+	±	-
50	19	10	0	0
50	25	7	1	2
50	29	9	0	1

[a] + most nymphs aggregated on the conditioned paper;
± about half of the nymphs aggregated on the conditioned paper and half on the control, or, almost none of them aggregated on either the conditioned or the control paper;
- most of nymphs aggregated on the control.

VI. Discussion

The ecological and physiological aspects of gregariousness in cockroaches have been studied extensively. Wille (1920) observed that nymphs of B. germanica remained almost constantly in groups during the first and second instars, but less so during the third instar. At usual room temperatures the older nymphs and adults lived completely isolated, but at certain temperatures they gathered together in large, tightly pressed groups.

Landowski (1938) studied the physiological or psychological effects of the gregariousness of Blatta orientalis L. on development and growth. He kept nymphs alone and in groups of 2, 4, 8, and 16 in jars of identical size and shape, and found that (1) mortality increased with the size of the group and with age, as each animal occupied more of the available space; (2) life in complete isolation extended the time required to produce an adult insect; and (3) the mean weight of the adult insect was, generally, in inverse proportion to the number of nymphs raised together; isolated insects usually attained the greatest adult weight.

Pettit (1940) studied gregariousness of the German cockroach and found that nymphs reared in isolation take a longer period for their maturation than those reared in groups. He considered that the altered metabolism under conditions of crowding and the resulting stimulation appeared merely to affect the time required to reach maturity. Ledoux (1945) studied gregariousness and social inter-attraction experimentally in B. orientalis and B. germanica. He concluded that group formation is not the result of chance, but is a social phenomenon, and that inter-attraction is mainly olfactory, conditioned by (1) positive chemotaxis to odors emitted by the cockroaches themselves; (2) positive hygrotaxis; and (3) thigmotaxis.

Willis et al. (1958) have also found that B. germanica, B. orientalis, and P. americana complete nymphal development in less time when reared in groups rather than individually. Recently, Wharton et al. (1967, 1968) studied precisely the effect of population density on growth and development of P. americana. They found that isolated nymphs grow more slowly than nymphs in groups, but the resultant adults were larger than those from groups of nymphs. When single nymphs were grown with adults, they grew as rapidly as nymphs in groups. This growth-promotion was attributed to physical contact with other nymphs or with adults.

The results of our current rearing experiments confirmed the previous results that growth and development of the German cockroach was accelerated when the nymphs were reared in groups of even 2 individuals, but the body-weight of the newly emerged adults reared in groups was almost the same as of those reared in isolation.

This group effect on the acceleration of growth was found when a young nymph of the German cockroach was reared in a group with an adult of the same species, or with a nymph of other species of cockroaches such as P. americana and P. fuliginosa (Serville) (Izutsu et al., 1968). The mechanism of the growth acceleration is being studied from ecological and physiological viewpoints.

104

The results of the current experiments indicate that chemical stimuli play an important role in aggregation, and that the chemical substance(s) responsible for aggregation is present on both the surface of the body and in the faeces. This suggests the following possibilities as to the origin of the substance responsible for aggregation: (1) 2 different substances from 2 different sources, 1 secreted from the region of the anus externally and deposited on the cuticular layer, and the other secreted from glands close to the anus internally and deposited on the faeces; (2) 1 substance secreted from the region of the anus externally and deposited on the faeces; (3) 1 substance secreted from glands close to the anus internally and deposited on the cuticular layer of the body.

An examination of the morphology of the rectum revealed the presence of 6 characteristic rectal pads in the anterior part of the rectum. According to Snodgrass (1935), the function of the pads is not definitely known and the supposed glandular nature of the organs has never been demonstrated. Wigglesworth (1932) has proposed a theory that the organs reabsorb water from the faecal matter in the rectum and thus play an important part in water conservation. This theory was experimentally confirmed in vitro in P. americana (Wall, 1967). Histological observations showed that the structure of cells surrounding the lumen of the rectum was quite different between the anterior and posterior sections. The structure of the rectal pad cells in the anterior part does not appear to be particularly adapted to absorption of water.

Several workers have found secretory areas for hormones and pheromones in the cells of the hind gut, e.g., the European corn borer, Ostrinia nubilalis (Hübner) (Beck and Alexander, 1964), the neotropical army ant, Eciton hamatum (F.) (Blum and Portocarrero, 1964). The actual localization of the site of sex pheromone production in bark beetles still remains obscure, but the hind gut seems to be concerned in the secretion (Wood, 1962; Vité et al., 1963; Pitman and Vité, 1963; Pitman et al., 1965; Wood et al., 1966; Zethner-Møller and Rudinsky, 1967).

In the present experiments, ether extracts of cockroach food did not elicit an aggregation response. The products of digestion still remain a possibility. However, filter papers conditioned with cockroaches which had had their abdominal tips removed between the eighth and ninth abdominal segments did not elicit the aggregation response.

105

Moreover, faeces excreted from cockroaches which had been
reared on different foods elicit an aggregation response.
  The active principle responsible for the aggregation
of the German cockroach seems to be a pheromone, and we pro-
pose to call this an "aggregation pheromone".
  It is suggested that the aggregation pheromone is produced in the rectal pad cells and secreted into the lumen
when faecal material is passing through the rectum. The
active principle found on the surface of terminal abdominal
segments is probably of the same origin, and is possibly
absorbed into the cuticular layer in the region of the anus
when faeces are excreted. Also, the pheromone could be
absorbed onto the lipid body surface after excretion.
  Isolation of the active principle is now in progress.

VII. Summary

  Young nymphs of the German cockroach, B. germanica,
aggregate. Newly hatched nymphs reared individually were
delayed in their growth and development. The facts that
aggregation was observed in darkness and that aggregation
does not occur in cockroaches with the antennae amputated
suggest that chemical stimuli initiate aggregation behavior.
  The active principle(s) responsible for aggregation was
found in faeces and in ether washings of the body surface,
especially the posterior portion of the abdomen. Histological study showed that the epithelium of the rectum formed
6 rectal pads consisting of a single layer of cells showing
a glandular structure. Filter papers conditioned with cockroaches from which the abdominal tips had been amputated did
not elicit the response for aggregation, while those conditioned with a methanol extract of the dissected rectums did.
The active principle is probably secreted from the rectal
pad cells into the lumen of the rectum, and then excreted
with the faeces. The active principle found on the body
surface was considered to be of the same origin as that
found in the faeces, and is absorbed by the lipids on the
body surface. This attractant is considered to be a pheromone.

VIII. References

Beck, S. D., and Alexander, N. (1964). Science 143, 478.
Blum, M. S., and Portocarrero, C. A. (1964). Ann. Ent. Soc. Amer. 57, 793.
Gould, G. E., and Deay, H. O. (1938). Ann. Ent. Soc. Amer. 31, 489.
Ishii, S., and Kuwabara, Y. (1967). Appl. Ent. & Zool. 2, 203.
Izutsu, M., Ueda, S., and Ishii, S. (1968). Annual Meeting Ent. Zool. Japan, Abstract, 27.
Landowski, J. (1938). Biol. Zentralblatt 58, 512.
Ledoux, A. (1945). Ann. Sci. Nat. Zool. 11, Sér. 7, 75.
Pettit, L. C. (1940). Entom. News 51, 293.
Pitman, G. B., and Vité, J. P. (1963). Contrib. Boyce Thomps. Inst. 22, 221.
Pitman, G. B., Kliefoth, R. A., and Vité, J. P. (1965). Contrib. Boyce Thomps. Inst. 23, 13.
Roth, L. M., and Willis, E. R. (1960). Smithsonian Misc. Coll. 141, 1.
Snodgrass, R. E. (1935). "Principles of Insect Morphology". McGraw-Hill Book Company, Inc. New York and London.
Vité, J. P., Gara, R. I., and Kliefoth, R. A. (1963). Contrib. Boyce Thomps. Inst. 22, 39.
Wall, B. J. (1967). J. Insect Physiol. 13, 565.
Wharton, D. R. A., Lola, J. E., and Wharton, M. L. (1967). J. Insect Physiol. 13, 699.
Wharton, D. R. A., Lola, J. E., and Wharton, M. L. (1968). J. Insect Physiol. 14, 637.
Wigglesworth, V. B. (1932). Quart. J. Micros. Sci. 75, 131.
Willis, E. R., Riser, G. R., and Roth, L. M. (1958). Ann. Ent. Soc. Amer. 51, 53.
Wille, J. (1920). Monogr. Angew. Entom. Nr. 5, Zeit. Angew. Entom., Beiheft 1, 7, 1.
Wood, D. L. (1962). Pan-Pacif. Entom. 37, 141.
Wood, D. L., Browne, L. E., Silverstein, R. M., and Rodin, J. O. (1966). J. Insect Physiol. 12, 523.
Zethner-Møller, O., and Rudinsky, J. A. (1967). J. Econ. Entom. 60, 575.

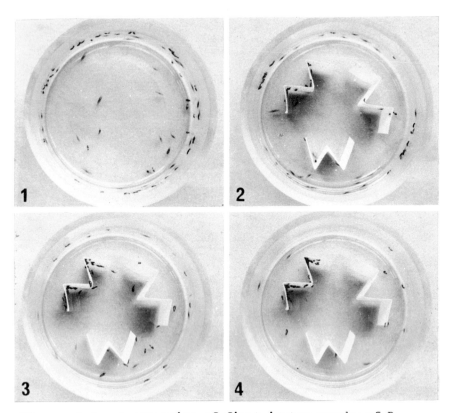

Figures 1-4.--Aggregation of first instar nymphs of B. germanica. 1. Approximately 60 nymphs were introduced into a glass pot. 2-4. Filter paper conditioned with cockroaches was placed at upper left. 2. After 1 min. 3. After 25 min. 4. After 43 min.

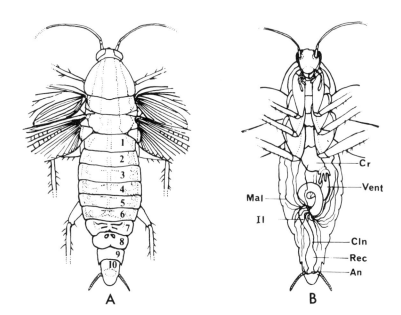

Figure 5.--Diagram showing the alimentary canal of the male
B. germanica in relation to abdominal segments. A.
Dorsal view. B. Ventral view. An: anus; Cr: crop;
Cln: colon; Il: ileum; Mal: Malpighian tubules;
Rec: rectum; Vent: ventriculus.

METHODOLOGY FOR ISOLATION, IDENTIFICATION
AND SYNTHESIS OF SEX PHEROMONES IN THE
LEPIDOPTERA

Martin Jacobson

Entomology Research Division
U.S. Department of Agriculture
Beltsville, Maryland

Table of Contents

Table of Contents

CONTROL OF INSECT BEHAVIOR

I. Gypsy moth (Porthetria dispar L.)

A. Isolation

The early chemical studies on this problem were conducted by Haller et al. (1944) and by Acree (1953, 1954),
who determined that the attractant was a lipid constituent
(or constituents) of the unsaponifiable neutral fraction of
a benzene extract of the last 2 abdominal segments of virgin females. However, it was not until several years later
that the sex attractants of this insect were isolated in
pure form by Jacobson et al. (1960, 1961). Chromatography
of the unsaponifiable fraction from 200,000 female moth
abdomens on successive columns of magnesium carbonate and
magnesium oxide by the method of Acree (1953), using ab-
sorbancy ratios at 249 and 285 mμ, yielded a yellow oil
highly attractive to male moths.

Subsequent investigations (Jacobson et al., 1961)
showed that these time-consuming operations (at least 3
days per column per 20,000 tips) could be avoided and
approximately equal fractionation obtained by repeatedly
dissolving the crude unsaponifiable fraction from 300,000
tips in acetone and filtering off the large amount of inactive solid, consisting of pigments, cholesterol, and
C_{27-31} hydrocarbons, that separated at room temperature, 5°
C, and -5° C. This procedure gave a total of 75 mg of active oil from 500,000 female moths. Ascending reversed-
phase chromatography on filter paper sheets impregnated with
silicone oil or polyethylene, using methanol-benzene-water
(5:1:1) as developing solvent, gave 5 spots having R_f values
of 0.0, 0.2, 0.34, 0.73, and 1.0, respectively. Only the
spot remaining at the point of origin was attractive to
males, and successive extraction of this zone with cold 95%
ethanol and with Skellysolve B gave 3.4 mg of white, waxy
crystals, mp 37.0-37.5° C, soluble in ethanol, and 20 mg of
colorless, Skellysolve-soluble liquid. These compounds
were attractive to male moths in the field at 10^{-2} and 10^{-7}
μg/trap, respectively.

B. Identification

Because of the small amount and low order of attractiveness of the solid component, only the major (liquid)
component was characterized. Gas chromatography showed it

to be pure and analysis showed its formula to be $C_{18}H_{34}O_3$. The infrared spectrum (Fig. 1) showed primary hydroxyl (3580, 3450, and 1042 cm^{-1}), ester carbonyl (1740 cm^{-1}) earmarked as acetoxyl by the strong band at 1234 cm^{-1}, since formates and higher esters absorb at considerably lower frequencies, a cis double bond (1660 and 783 cm^{-1}), and an unbroken chain of at least 4 methylene groups (720 cm^{-1}). The attractant possessed optical activity, $[\alpha]_D^5$ + 7.9° C.

Hydrogenation of 2 mg of the attractant in a specially-designed micro-hydrogenator, shown in Fig. 2, showed the presence of only one double bond and a molecular weight of 298. Platinum oxide catalyst in 0.5 ml of absolute ethanol was first reduced in the reaction vessel by passing through hydrogen from the graduated tube with the stopcock in position B. The weighed sample in a small aluminum boat fitting into the glass-jointed spoon was then introduced directly by merely inverting the spoon, thus dropping the boat and sample into the liquid. The amount of hydrogen necessary to hydrogenate the sample was measured by means of the mercury leveling bulb, and the apparatus was opened to the atmosphere by turning the stopcock to position A. Saponification of the saturated compound with alcoholic alkali was unsuccessful but went smoothly in 3 min at 120-125° using potassium hydroxide in diethylene glycol, conditions found to be sufficient for the saponification of esters of most secondary alcohols. The saponification products were acetic acid and a crystalline diol.

Lack of absorption in the ultraviolet precluded the 2-position for the attractant double bond, and its optical activity showed that the acetoxyl group did not lie on an unsaturated carbon atom, since an asymmetric carbon atom would not then exist. In addition, stability of the attractant to high-temperature vapor phase chromatography on Craig polyester succinate indicated the probability that at least one methylene group separated the acetoxyl group from the double bond.

Micro-oxidation of the attractant with periodate-permanganate reagent, by the method of von Rudloff (1956), cleaved the double bond only, leaving the hydroxyl and acetoxyl groups intact. A 92% yield of 3-acetoxy-1-nonanoic acid was obtained, together with an ω-hydroxy acid which was further oxidized in 71% yield with alkaline permanganate to pimelic acid.

Thus, the only structure for the attractant, named

"gyptol", consistent with the foregoing data was (+)-10-acetoxy-cis-7-hexadecen-1-ol (structure I).

$$CH_3(CH_2)_5CHCH_2CH\overset{c}{=}CH(CH_2)_6OH$$

$$OCOCH_3$$

(I)

C. Synthesis

The dl (racemic)-form of (I) was synthesized in 0.2% over-all yield by the following scheme (Jacobson et al., 1961):

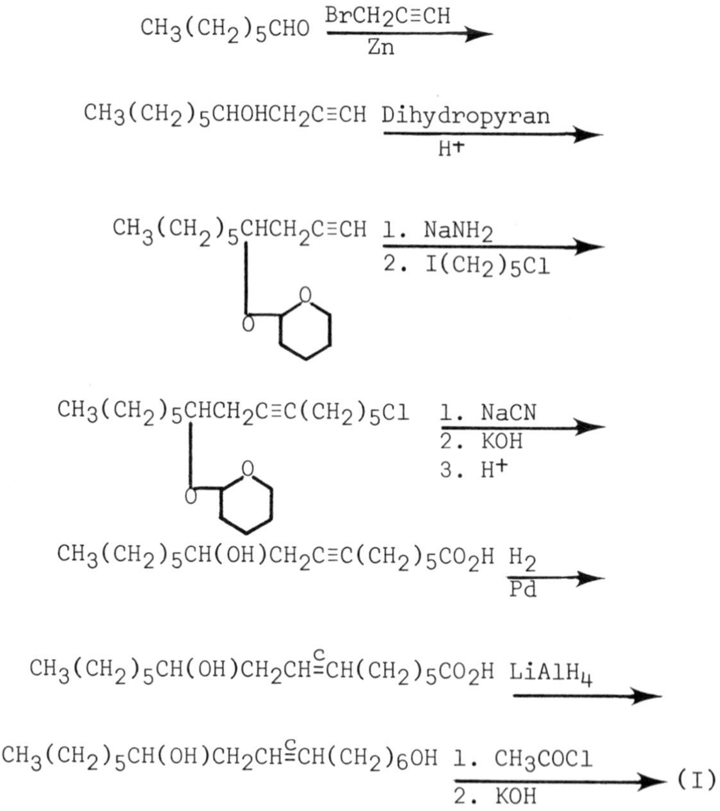

$$CH_3(CH_2)_5CHO \xrightarrow[Zn]{BrCH_2C\equiv CH}$$

$$CH_3(CH_2)_5CHOHCH_2C\equiv CH \xrightarrow[H^+]{Dihydropyran}$$

$$CH_3(CH_2)_5CHCH_2C\equiv CH \xrightarrow[2.\ I(CH_2)_5Cl]{1.\ NaNH_2}$$

$$CH_3(CH_2)_5CHCH_2C\equiv C(CH_2)_5Cl \xrightarrow[\substack{2.\ KOH \\ 3.\ H^+}]{1.\ NaCN}$$

$$CH_3(CH_2)_5CH(OH)CH_2C\equiv C(CH_2)_5CO_2H \xrightarrow[Pd]{H_2}$$

$$CH_3(CH_2)_5CH(OH)CH_2CH\overset{c}{=}CH(CH_2)_5CO_2H \xrightarrow{LiAlH_4}$$

$$CH_3(CH_2)_5CH(OH)CH_2CH\overset{c}{=}CH(CH_2)_6OH \xrightarrow[2.\ KOH]{1.\ CH_3COCl} (I)$$

It was identical in all respects save optical activity with the natural attractant. This form was successfully re-solved into the d- and l-forms by treating its acid succinate with L-brucine, separating the brucine salts by fractional crystallization from acetone, decomposing the salts, and saponifying the acid succinates with ethanolic alkali (Jacobson, 1962).

Although the synthesis of racemic gyptol has also been reported by Eiter et al. (1967), the infrared spectrum of their product showed certain key differences from that of Jacobson et al. (1961), and it is not definitely known whether the two products are the same.

II. Silkworm moth (Bombyx mori (L.))

A. Isolation

Impure attractant was obtained by macerating virgin female abdomens with benzene (Butenandt, 1941, 1955) and macerating the abdomens of mated females with ethanol (Makino et al., 1956). Hecker (1956) reported that the pure 4-(p-nitrophenylazo)-benzoate of the attractant had been prepared; saponification of this ester gave a substance showing unusually low activity. However, he proposed that the attractant was a doubly conjugated alcohol of 12-15 carbon atoms.

In 1959 (Butenandt et al., 1959), the pure attractant, designated "bombykol", was obtained as its 4'-nitroazoben-zenecarboxylic acid ester by extracting 500,000 virgin female abdominal tips with ethanol-ether (3:1), saponifying the extract with alkali, freeing the neutral fraction of cholesterol, esterifying with succinic anhydride, saponify-ing the succinates, and treating with 4'-nitroazobenzene-carboxylic acid chloride. Digestion of the resulting ester with petroleum ether gave 3 fractions which were chromato-graphed 3 times on hydrophobic silica gel, using paraffin oil as the stationary phase and 85 and 90% acetone as the mobile phases, to yield 12 mg of derivative, mp 95-96° C (Butenandt et al., 1961a). Saponification of the derivative regenerated the pure attractant.

B. Identification

The attractant derivative showed a molecular formula of $C_{29}H_{37}N_3O_4$ and a molecular weight of 491.6, indicating

that the pure attractant was a doubly-unsaturated alcohol
having the formula $C_{16}H_{30}O$. Ultraviolet absorption at 230
mμ for the derivative showed that the double bonds were
conjugated (Butenandt et al., 1961b).

The infrared spectrum of free bombykol showed it to
possess a primary hydroxyl group and 2 conjugated double
bonds indicative of cis,trans unsaturation. Catalytic
hydrogenation with platinum oxide gave palmityl alcohol,
and permanganate oxidation of the derivative yielded acid
moieties which were identified by paper chromatography of
their methyl esters. By means of this brilliant investigation, bombykol was shown to be 10,12-hexadecadien-1-ol
(structure II) (Butenandt et al., 1961b).

$$CH_3(CH_2)_2CH=CHCH=CH(CH_2)_9OH$$

(II)

Although the geometrical configuration of bombykol had
at first been thought to be cis,trans (Butenandt et al.,
1959), subsequent synthesis (see IIC) of the 4 possible
isomers of (II) showed it to possess the trans-10,cis-12
form (Butenandt et al., 1962; Hecker, 1960; Truscheit and
Eiter, 1962).

C. Synthesis

The following synthetic schemes have been used to prepare bombykol (Butenandt et al., 1962):

$$CH_3(CH_2)_2Br \xrightarrow{NaC\equiv CH} CH_3(CH_2)_2C\equiv CH \xrightarrow[HCHO]{Mg}$$

$$CH_3(CH_2)_2C\equiv CCH_2OH \xrightarrow{PBr_3} CH_3(CH_2)_2C\equiv CCH_2Br \xrightarrow{Ph_3P}$$

$$CH_3(CH_2)_2C\equiv CCH_2P^+Ph_3Br^- \xrightarrow[EtONa/EtOH]{OHC(CH_2)_8CO_2Et}$$

$$CH_3(CH_2)_2C\equiv CCH=CH(CH_2)_8CO_2Et \xrightarrow{urea}$$

(cis + trans)

117

$$CH_3(CH_2)_2C \equiv CCH \overset{t}{=} CH(CH_2)_8CO_2Et \qquad H_2, Pd \longrightarrow$$

$$CH_3(CH_2)_2CH \overset{c}{=} CHCH \overset{t}{=} CH(CH_2)_8CO_2Et \qquad LiAlH_4 \longrightarrow \qquad (II)$$

(Truscheit and Eiter, 1962):

$$CH_3OCO(CH_2)_9CHO \qquad \frac{BrCH_2C \equiv CH}{Zn} \longrightarrow$$

$$CH_3OCO(CH_2)_9C(OH)HCH_2C \equiv CH \qquad P_2O_5 \longrightarrow$$

$$CH_3OCO(CH_2)_9CH \overset{t}{=} CHC \equiv CH \qquad \begin{array}{l} 1.\ OH^- \\ \overline{2.\ Mg} \\ 3.\ CH_3(CH_2)_2Br \end{array} \longrightarrow$$

$$CH_3(CH_2)_2C \equiv CCH \overset{t}{=} CH(CH_2)_9OH \qquad H_2, Pd \longrightarrow \qquad (II)$$

III. Pink bollworm moth
(Pectinophora gossypiella (Saunders))

A. Isolation

The whole bodies of 850,000 virgin female moths were macerated in a blender with methylene chloride, the extract was freed of a considerable amount of inactive solids by crystallization from 10 volumes of acetone at -20° C, the filtrate was freed of solvent and partitioned into cold methanol, and the methanol-soluble fraction was again crystallized from acetone at -20° C. Column chromatography of the soluble oil on 2 columns of Florisil, and then by chromatography of the 3% ether-hexane eluate fraction on 2 columns of silica gel impregnated with 25% silver nitrate gave active fractions that were eluted with 25 and 50% ether-hexane. Preparative gas chromatography of these active fractions on 5% SE-30 (Chromosorb W base) at 185° C gave 1.6 mg of the pure sex pheromone (Jones et al., 1966).

The pheromone (designated "propylure") proved to be

highly exciting sexually to male pink bollworm moths in the laboratory (Jones et al., 1966) but did not attract males to field traps (Jones and Jacobson, 1968). However, mixtures of propylure with N,N-diethyl-m-toluamide ("deet"), isolated from the methanol-insoluble portion of the female extract by chromatography on silver nitrate-impregnated silicic acid, did attract males to such traps (Jones and Jacobson, 1968).

B. Identification

   The infrared spectrum of propylure showed strong bands at 1755 and 1235 cm^{-1} and a medium band at 1038 cm^{-1}, characteristic of a primary acetate group, as well as bands at 1660 (unsaturation), 965 (trans), and 723 cm^{-1} (at least 4 successive methylene groups). Hydrogenolytic gas chromatography (Beroza and Sarmiento, 1963) established the presence of branching in the molecule, and the mass spectrum gave the molecular formula as $C_{18}H_{32}O_2$ (molecular weight 280). These data characterized propylure as the acetate of a branched-chain, C_{16}, primary alcohol with 2 double bonds, which could not be conjugated because of the absence of ultraviolet absorption. Nuclear magnetic resonance spectra gave evidence for 3 olefinic protons, 2 methylene protons adjacent to an acetate oxygen atom, 10 methylene protons adjacent to double-bonded carbon atoms, 3 acetyl methyl protons, 8 methylene protons adjacent to either methylene or methyl groups, and 6 terminal methyl protons separated from a double-bonded carbon by at least 2 methylene groups. Double resonance studies indicated at least 3 methylene groups between the acetate group and a double bond, and only 2 methylene groups between the double bonds (Jones et al., 1966).
   The only structure for propylure consistent with the foregoing data was 10-propyl-trans-5,9-tridecadienyl acetate (structure III).

$$CH_3CH_2CH_2$$
$$CH_3CH_2CH_2 \Big\rangle C = CH(CH_2)_2CH\overset{t}{=}CH(CH_2)_4OCOCH_3$$

(III)

119

C. Synthesis

Propylure was synthesized in 0.2% overall yield by Jones et al. (1966) according to the following scheme:

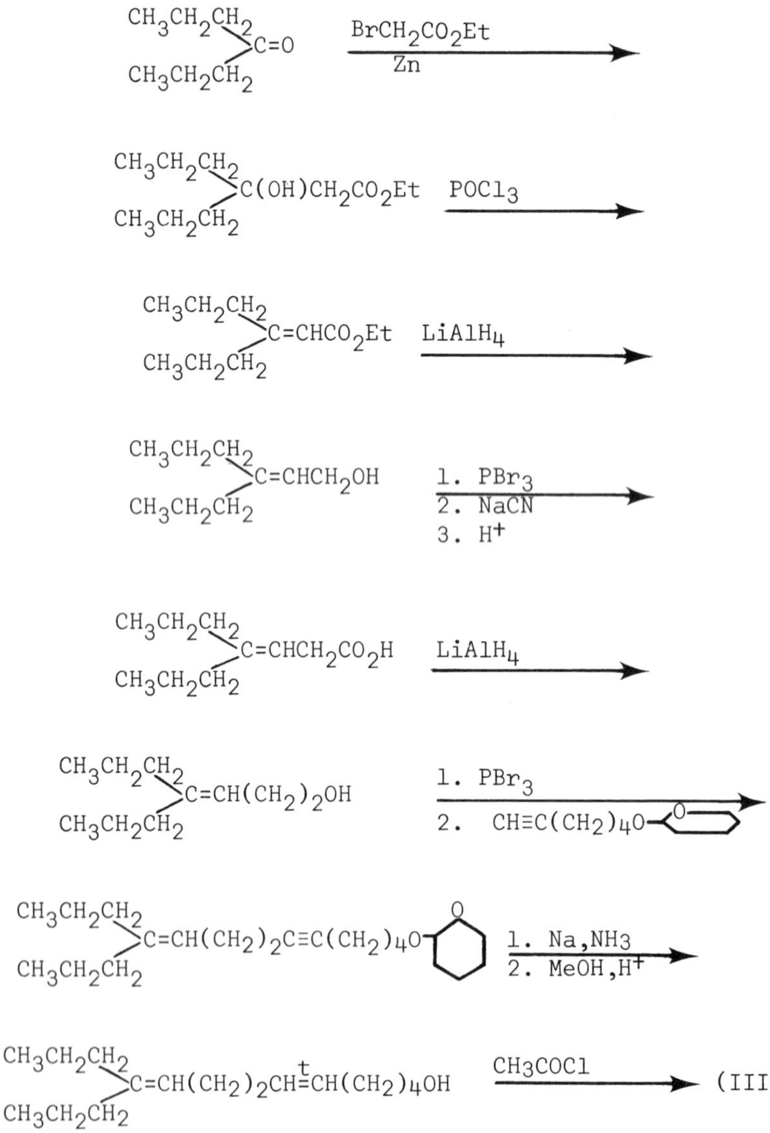

$$\begin{array}{c} CH_3CH_2CH_2 \\ CH_3CH_2CH_2 \end{array} \!\! C=O \quad \xrightarrow[\text{Zn}]{BrCH_2CO_2Et}$$

$$\begin{array}{c} CH_3CH_2CH_2 \\ CH_3CH_2CH_2 \end{array} \!\! C(OH)CH_2CO_2Et \quad POCl_3 \quad \longrightarrow$$

$$\begin{array}{c} CH_3CH_2CH_2 \\ CH_3CH_2CH_2 \end{array} \!\! C=CHCO_2Et \quad LiAlH_4 \quad \longrightarrow$$

$$\begin{array}{c} CH_3CH_2CH_2 \\ CH_3CH_2CH_2 \end{array} \!\! C=CHCH_2OH \quad \begin{array}{l} 1.\ PBr_3 \\ 2.\ NaCN \\ 3.\ H^+ \end{array} \longrightarrow$$

$$\begin{array}{c} CH_3CH_2CH_2 \\ CH_3CH_2CH_2 \end{array} \!\! C=CHCH_2CO_2H \quad LiAlH_4 \quad \longrightarrow$$

$$\begin{array}{c} CH_3CH_2CH_2 \\ CH_3CH_2CH_2 \end{array} \!\! C=CH(CH_2)_2OH \quad \begin{array}{l} 1.\ PBr_3 \\ 2.\ CH\equiv C(CH_2)_4O- \end{array}$$

$$\begin{array}{c} CH_3CH_2CH_2 \\ CH_3CH_2CH_2 \end{array} \!\! C=CH(CH_2)_2C\equiv C(CH_2)_4O- \quad \begin{array}{l} 1.\ Na,NH_3 \\ 2.\ MeOH,H^+ \end{array} \longrightarrow$$

$$\begin{array}{c} CH_3CH_2CH_2 \\ CH_3CH_2CH_2 \end{array} \!\! C=CH(CH_2)_2CH\overset{t}{=}CH(CH_2)_4OH \quad \xrightarrow{CH_3COCl} \quad (III)$$

Eiter et al. (1967) also synthesized propylure according to the following scheme:

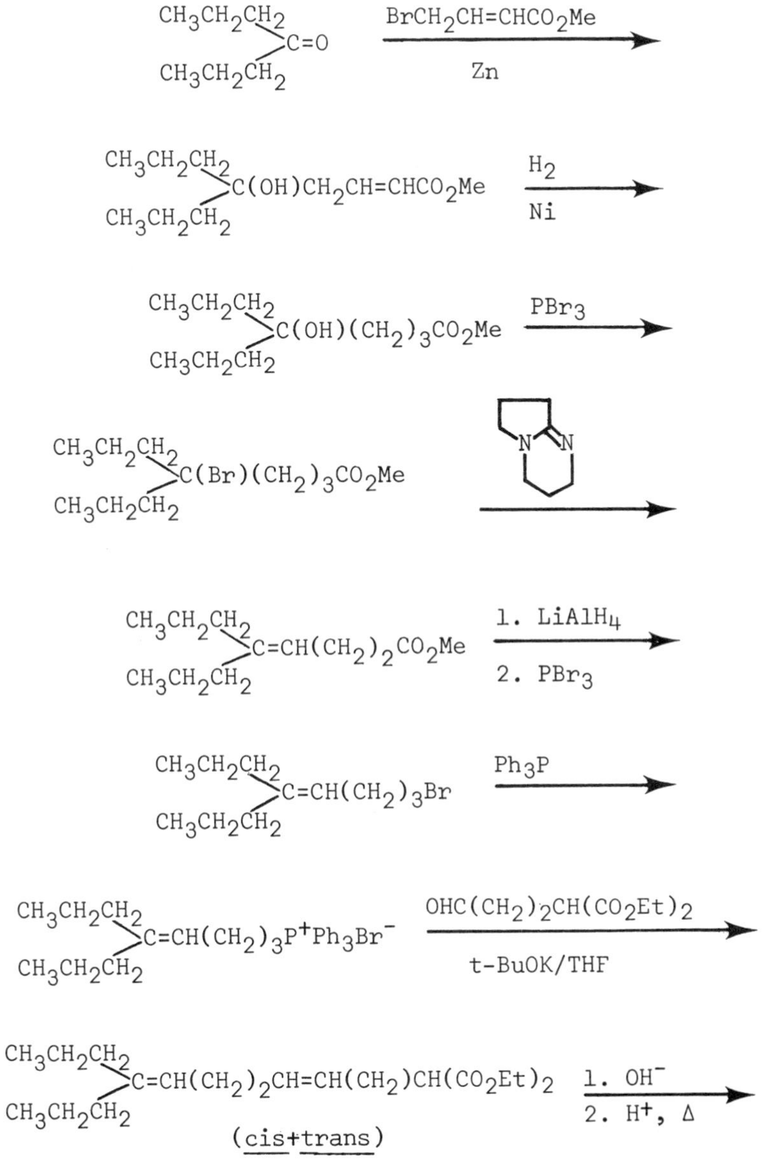

$$\begin{array}{c} CH_3CH_2CH_2 \\ \diagdown \\ CH_3CH_2CH_2 \end{array} C=O \quad \xrightarrow[\text{Zn}]{\text{BrCH}_2\text{CH=CHCO}_2\text{Me}}$$

$$\begin{array}{c} CH_3CH_2CH_2 \\ \diagdown \\ CH_3CH_2CH_2 \end{array} C(OH)CH_2CH=CHCO_2Me \quad \xrightarrow[\text{Ni}]{\text{H}_2}$$

$$\begin{array}{c} CH_3CH_2CH_2 \\ \diagdown \\ CH_3CH_2CH_2 \end{array} C(OH)(CH_2)_3CO_2Me \quad \xrightarrow{\text{PBr}_3}$$

$$\begin{array}{c} CH_3CH_2CH_2 \\ \diagdown \\ CH_3CH_2CH_2 \end{array} C(Br)(CH_2)_3CO_2Me \quad \xrightarrow{\quad\quad}$$

$$\begin{array}{c} CH_3CH_2CH_2 \\ \diagdown \\ CH_3CH_2CH_2 \end{array} C=CH(CH_2)_2CO_2Me \quad \xrightarrow[\text{2. PBr}_3]{\text{1. LiAlH}_4}$$

$$\begin{array}{c} CH_3CH_2CH_2 \\ \diagdown \\ CH_3CH_2CH_2 \end{array} C=CH(CH_2)_3Br \quad \xrightarrow{\text{Ph}_3\text{P}}$$

$$\begin{array}{c} CH_3CH_2CH_2 \\ \diagdown \\ CH_3CH_2CH_2 \end{array} C=CH(CH_2)_3P^+Ph_3Br^- \quad \xrightarrow[\text{t-BuOK/THF}]{\text{OHC(CH}_2)_2\text{CH(CO}_2\text{Et)}_2}$$

$$\begin{array}{c} CH_3CH_2CH_2 \\ \diagdown \\ CH_3CH_2CH_2 \end{array} C=CH(CH_2)_2CH=CH(CH_2)CH(CO_2Et)_2 \quad \xrightarrow[\text{2. H}^+, \Delta]{\text{1. OH}^-}$$

(cis+trans)

121

$$\begin{array}{c} CH_3CH_2CH_2 \\ {>}C{=}CH(CH_2)_2CH{=}CH(CH_2)_3CO_2H \\ CH_3CH_2CH_2 \end{array} \xrightarrow[\text{2. } CH_3COCl]{\text{1. } LiAlH_4} (III)$$

(cis+trans)

IV.   Cabbage looper moth (Trichoplusia ni (Hübner))

A.   Isolation

An extract prepared by grinding the last 2-3 abdominal segments of virgin female moths with methylene chloride in a blender was chromatographed on silicic acid (Berger et al., 1964) and the active fractions were combined and purified by gas chromatography on 12% diethylene glycol succinate (Anakrom ABS base) at 172° C (Berger, 1966).

B.   Identification

The infrared spectrum of the attractant indicated that it was a long-chain acetate (1740, 1240, and 720 cm^{-1}), and its unsaturation was determined by bromine addition. Saponification gave an alcohol moiety shown to be primary by infrared absorption (3300 and 1060 cm^{-1}). Hydrogenation of this alcohol showed the presence of one double bond and gave dodecanol, identified by gas chromatography of its acetate ester.

Absence of a band at 980-965 cm^{-1} in the infrared spectrum showed that the double bond did not have the trans configuration. The position of the double bond in the alcohol was located between C-7 and C-8 by ozonolysis, which yielded amyl alcohol and 1,7-heptanediol.

The foregoing data indicated that the attractant was cis-7-dodecen-1-ol acetate (structure IV), and this was confirmed by synthesis (Berger, 1966).

$$CH_3(CH_2)_3CH\overset{c}{=}CH(CH_2)_6OCOCH_3$$

(IV)

C.   Synthesis

Berger (1966) synthesized the attractant in an overall yield of 22% by the following scheme:

122

$$CH_3(CH_2)_3C \equiv CH \quad \xrightarrow[\text{2. } I(CH_2)_5Cl]{\text{1. } Na,NH_3}$$

$$CH_3(CH_2)_3C \equiv C(CH_2)_5Cl \quad \xrightarrow[\substack{\text{2. } OH^- \\ \text{3. } H^+}]{\text{1. } NaCN}$$

$$CH_3(CH_2)_3C \equiv C(CH_2)_5CO_2H \quad \xrightarrow[Pd]{H_2}$$

$$CH_3(CH_2)_3\overset{c}{CH} = CH(CH_2)_5CO_2H \quad \xrightarrow{LiAlH_4}$$

$$CH_3(CH_2)_3\overset{c}{CH} = CH(CH_2)_6OH \quad \xrightarrow{CH_3COCl} \quad (IV)$$

A more direct procedure was used by Green et al. (1967):

$$CH_3(CH_2)_3C \equiv C(CH_2)_6OCOCH_3 \quad \xrightarrow[Pd]{H_2} \quad (IV)$$

An overall yield of at least 37% is attainable and the method has been adapted to commercial production of the attractant.

MARTIN JACOBSON

V. Fall armyworm moth (Spodoptera frugiperda (J. E. Smith))

A. Isolation

Abdomens of virgin female moths were homogenized in a
blender with ether, the extract was subjected to low-temper-
ature (-20° C) acetone crystallization, and the acetone-
soluble fraction was chromatographed on a column
of silicic acid, eluting successively with pentane and
increasing percentages of ether in pentane. Activity was
recovered in the 1% and 3% ether-pentane fractions, and
these were combined and shaken repeatedly with portions of
methanol. The methanol-soluble fraction was further purified by successive thin-layer chromatography on silica gel
G, column chromatography on silicic acid, thin-layer chromatography on acetylated cellulose, and preparative gas
chromatography on 10% SE-30 (Chromosorb P base) at 187° C.
In this way, a total of 900 µg of pure pheromone was isolated from 135,000 female moths (Sekul and Sparks, 1967).

B. Identification

The infrared spectrum of the pheromone exhibited
strong sharp bands at 1735 and 1235 cm^{-1} (acetate ester)
and bands at 1650 and 720 (cis unsaturation). Absence of
ultraviolet absorption precluded conjugation. Alkaline
saponification destroyed activity and acetylation of the
product restored activity. Microhydrogenation of the pheromone gave tetradecanol acetate, indicating that the pheromone was an unsaturated straight-chain 16-carbon acetate
(acetate of a cis-double bonded, primary, 14-carbon alcohol).
The position of unsaturation was determined by ozonolysis of the alcohol to be between C-9 and C-10, since the
ozonolytic fragmentation products were amyl alcohol and
1,9-nonanediol. The pheromone thus appeared to be cis-9-
tetradecen-1-ol acetate (structure V), and this was confirmed by synthesis (Sekul and Sparks, 1967).

$$CH_3(CH_2)_3CH\overset{c}{=}CH(CH_2)_8OCOCH_3$$

(V)

C. Synthesis

The pheromone was readily synthesized in 2 steps

124

(Sekul and Sparks, 1967) by reducing methyl myristoleate with lithium aluminum hydride to myristoleyl alcohol and acetylating this to give V. However, methyl myristoleate is extremely rare and therefore quite expensive. The procedure developed by Warthen (1968) is quite straightforward and commerically adaptable.

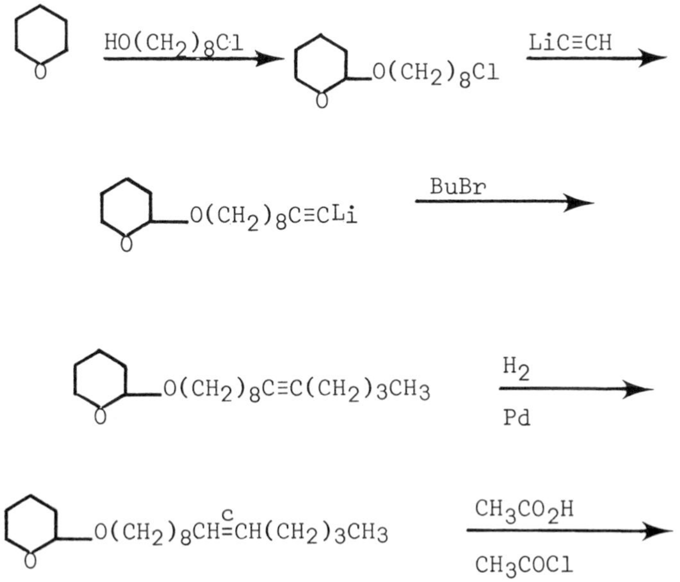

The pheromone has also been synthesized (Jacobson and Harding, 1968) by using the cabbage looper sex attractant (IV) as starting material.

$$CH_3(CH_2)_3CH\overset{c}{=}CH(CH_2)_6OCOCH_3 \quad \xrightarrow{\begin{array}{l}1.\ OH^-\\2.\ PBr_3\end{array}}$$

(IV)

$$CH_3(CH_2)_3CH\overset{c}{=}CH(CH_2)_6Br \quad \xrightarrow{CHNa(CO_2Et)_2}$$

125

$$CH_3(CH_2)_3CH\overset{c}{=}CH(CH_2)_6CH(CO_2Et)_2 \quad \xrightarrow[2.\ H^+]{1.\ OH^-}$$

$$CH_3(CH_2)_3CH\overset{c}{=}CH(CH_2)_6CH(CO_2H)_2 \quad \xrightarrow{\Delta}$$

$$CH_3(CH_2)_3CH\overset{c}{=}CH(CH_2)_7CO_2H \quad \xrightarrow{LiAlH_4}$$

$$CH_3(CH_2)_3CH=CH(CH_2)_8OH \quad \xrightarrow{CH_3COCl} \quad (V)$$

VI. Red-banded leaf roller moth
(Argyrotaenia velutinana (Walker))

A. Isolation

Preliminary investigations (Roelofs and Feng, 1967) showed that extracts of the last 2 abdominal segments of virgin female moths prepared with ether and methylene chloride were equally attractive to males, whereas extracts prepared with acetone, benzene, chloroform, methanol, or 95% ethanol were less attractive. Ether extracts could be cleaned up by column chromatography on silicic acid (active material was eluted with 5% ether-petroleum ether) or, preferably, on Florisil (active material was eluted with 15% ether-petroleum ether).

The best isolation procedure (Roelofs and Arn, 1968a, 1968b) proved to be as follows. The last two abdominal segments of 40,000 virgin females were mechanically pulverized with methylene chloride and the extract was chromatographed on successive columns of Florisil and silver nitrate-impregnated silicic acid, using petroleum ether and increasing percentages of ether in petroleum ether as eluting solvents. The active fractions were obtained with 15 and 5% ether-petroleum ether, respectively. Preparative gas chromatography of active material purified in this manner gave 200 μg of pure pheromone as single peaks from 3% cyclohexanedimethanol succinate (Chromosorb Q base) or

126

10% JXR packing (Chromosorb Q base).

B. Identification

Saponification of crude ether extracts of the moths
followed by reacetylation of the inactive saponification
products had yielded active material, whereas propionylation
did not restore activity. These were strong indications
that the attractant was an acetate. The extract rapidly
lost all activity following treatment with bromine at room
temperature, showing that the attractant was unsaturated
(Roelofs and Feng, 1967).
    The pure attractant collected by gas chromatography
showed a retention time similar to that calculated for a
mono-unsaturated 14-carbon-chain acetate, and catalytic
hydrogenation yielded a product with the retention time of
tetradecyl acetate. Infrared and nuclear magnetic resonance
spectra and ozonolysis of the pure attractant (propionaldehyde and 11-acetoxyundecanal were obtained as ozonolysis
products) showed that the attractant was 11-tetradecen-1-ol
acetate. Synthesis of the cis and trans isomers of this
compound showed the natural attractant to be the cis isomer,
and its structure is thus (VI) (Roelofs and Arn, 1968a,
1968b).

$$CH_3CH_2CH\overset{c}{=}CH(CH_2)_{10}OCOCH_3$$

(VI)

C. Synthesis

An economical synthesis of VI was achieved by means of
a novel one-pot reaction sequence, condensing 11-bromoun-
decyl acetate and propionaldehyde by a Wittig reaction,
using dimethylformamide as solvent and sodium methoxide as
catalyst. The following multiple-step synthesis was also
successful (Roelofs and Arn, 1968b):

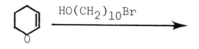

$$\text{O(CH}_2)_{10}\text{Br} \xrightarrow{\text{LiC} \equiv \text{CCH}_2\text{CH}_3}$$

$$\text{O(CH}_2)_{10}\text{C} \equiv \text{CCH}_2\text{CH}_3 \xrightarrow{\text{H}^+}$$

$$CH_3CH_2C \equiv C(CH_2)_{10}OH \xrightarrow[\text{2. CH}_3\text{COCl}]{\text{1. H}_2,\text{Pd}} \quad (VI)$$

VII. False codling moth (Argyroploce leucotreta Meyr.)

A. Isolation

A crude extract (solvent not given) of the female moths contained 4 major components, only one of which showed biological activity. This component was isolated by gas chromatography (Read et al., 1968).

B. Identification

Activity of the crude extract was destroyed by alkaline hydrolysis and by catalytic hydrogenation, but the activity of the hydrolyzed product was restored on acetylation. The pheromone, on gas chromatography over ethylene glycol adipate and methyl silicone gum rubber columns, showed a retention time midway between methyl dodecanoate and methyl tetradecanoate. These results were indicative of an unsaturated C_{12} straight-chain acetate. The retention times and mass spectrum of the pheromone were identical with those of synthetic trans-7-dodecen-1-ol acetate (VII), and oxidation of these compounds with periodate-permanganate gave identical fragments (Read et al., 1968).

$$CH_3(CH_2)_3\overset{t}{C}H=CH(CH_2)_6OCOCH_3$$

(VII)

C. Synthesis

Compound (VII) was synthesized (Green et al., 1967) by the following scheme:

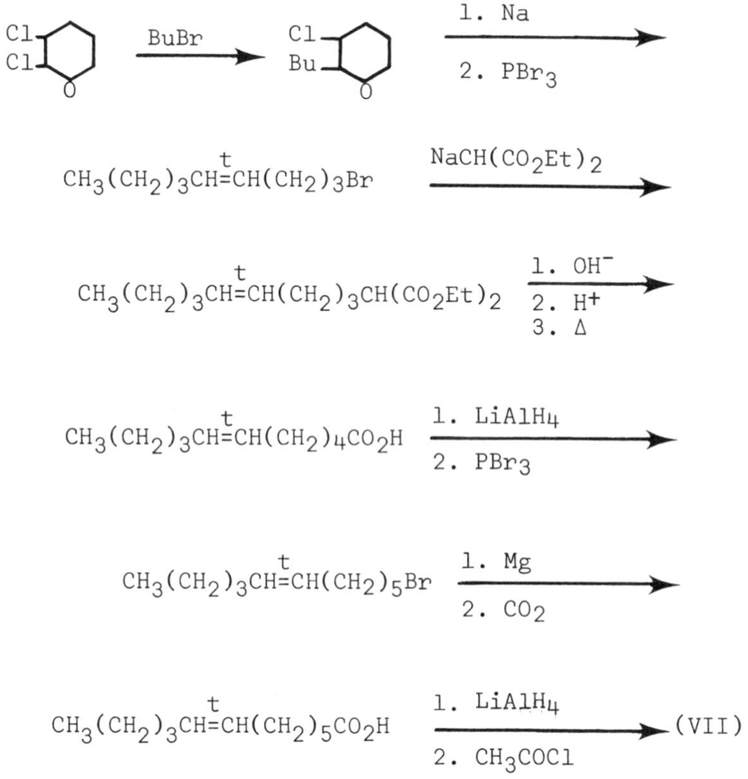

Cl, Cl cyclohexanone $\xrightarrow{\text{BuBr}}$ Cl, Bu cyclohexanone $\xrightarrow[\text{2. PBr}_3]{\text{1. Na}}$

$$CH_3(CH_2)_3\overset{t}{CH}=CH(CH_2)_3Br \xrightarrow{NaCH(CO_2Et)_2}$$

$$CH_3(CH_2)_3\overset{t}{CH}=CH(CH_2)_3CH(CO_2Et)_2 \xrightarrow[\substack{2.\ H^+ \\ 3.\ \Delta}]{1.\ OH^-}$$

$$CH_3(CH_2)_3\overset{t}{CH}=CH(CH_2)_4CO_2H \xrightarrow[\text{2. PBr}_3]{\text{1. LiAlH}_4}$$

$$CH_3(CH_2)_3\overset{t}{CH}=CH(CH_2)_5Br \xrightarrow[\text{2. CO}_2]{\text{1. Mg}}$$

$$CH_3(CH_2)_3\overset{t}{CH}=CH(CH_2)_5CO_2H \xrightarrow[\text{2. CH}_3COCl]{\text{1. LiAlH}_4} (VII)$$

VIII. European corn borer (Ostrinia nubilalis (Hübner))

A. Isolation

Virgin female moths were homogenized in a blender with 1,2-dichloroethane and the extract was clarified by precipitation of inactive lipids from acetone at -70° C and chromatographed on several successive columns of silicic acid. The active fractions were combined and the pure pheromone was isolated by preparative gas chromatography, first on 5% SE-30 (Chromosorb G base) at 200° and then on 5% diethylene glycol adipate (acid-washed, dimethylchlorosilane-treated Chromosorb W base) at 185° C (Klun, 1968).

B. Identification

The pheromone is a strong electron captor soluble in polar organic solvents, and nonsaponifiable. It thus appears to be an alcohol, but it has not yet been identified (Klun, 1968).

IX. References

Acree, F., Jr. (1953). J. Econ. Entomol. 46, 313, 900.
Acree, F., Jr. (1954). J. Econ. Entomol. 47, 321.
Berger, R. S. (1966). Ann. Entomol. Soc. Amer. 59, 767.
Berger, R. S., McGough, J. M., Martin, D. F., and Ball, L.
    R. (1964). Ann. Entomol. Soc. Amer. 57, 606.
Beroza, M., and Sarmiento, R. (1963). Anal. Chem. 35, 1353.
Butenandt, A. (1941). Angew. Chem. 54, 89.
Butenandt, A. (1955). Nova Acta Leopoldina 17, 445.
Butenandt, A., Beckmann, R., Stamm, D., and Hecker, E.
    (1959). Z. Naturforsch. 14b, 283.
Butenandt, A., Beckmann, R., and Hecker, E. (1961a). Z.
    Physiol. Chem., Hoppe-Seyler's 324, 71.
Butenandt, A., Beckmann, R., and Stamm, D. (1961b). Z.
    Physiol. Chem., Hoppe-Seyler's 324, 84.
Butenandt, A., Hecker, E., Hopp, M., and Koch, W. (1962).
    Justus Liebig's Ann. Chem. 658, 39.
Eiter, K., Truscheit, E., and Boness, M. (1967). Justus
    Liebig's Ann. Chem. 709, 29.
Green, N., Jacobson, M., Henneberry, T. J., and Kishaba,
    A. N. (1967). J. Med. Chem. 10, 533.
Haller, H. L., Acree, F., Jr., and Potts, S. F. (1944).
    J. Amer. Chem. Soc. 66, 1659.
Hecker, E. (1956). Proc. Xth Intern. Cong. Entomol. 2,
    293 (Pub. 1958).
Hecker, E. (1960). Proc. XIth Intern. Cong. Entomol. 3B,
    69 (Pub. 1961).
Jacobson, M. (1962). J. Org. Chem. 27, 2670.
Jacobson, M., Beroza, M., and Jones, W. A. (1960). Science
    132, 1011.
Jacobson, M., Beroza, M., and Jones, W. A. (1961). J.
    Amer. Chem. Soc. 83, 4819.
Jacobson, M., and Harding, C. (1968). J. Econ. Entomol. 61,
    394.
Jones, W. A., and Jacobson, M. (1968). Science 159, 99.
Jones, W. A., Jacobson, M., and Martin, D. F. (1966).
    Science 152, 1516.
Klun, J. A. (1968). J. Econ. Entomol. 61, 484.
Makino, K., Satoh, K., and Inagami, K. (1956). Biochem.
    Biophys. Acta 19, 394.
Read, J. S., Warren, F. L., and Hewitt, P. H. (1968). Chem.
    Commun. No. 14, 792.
Roelofs, W. L., and Arn, H. (1968a). New York's Food Life
    Sci. 1(1), 13.

MARTIN JACOBSON

Roelofs, W. L., and Arn, H. (1968b). Nature 219, 513.
Roelofs, W. L., and Feng, K.-C. (1967). Ann. Entomol. Soc.
  Amer. 60, 1199.
von Rudloff, E. (1956). J. Amer. Oil Chem. Soc. 33, 126.
Sekul, A. A., and Sparks, A. N. (1967). J. Econ. Entomol.
  60, 1270.
Truscheit, E., and Eiter, K. (1962). Justus Liebig's Ann.
  Chem. 658, 65.
Warthen, D. (1968). J. Med. Chem. 11, 371.

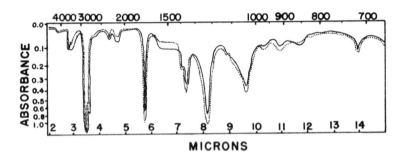

Figure 1.--Infrared spectrum of gypsy moth sex pheromone.

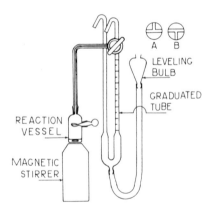

Figure 2.--Micro hydrogenation apparatus.

VOLATILE ORGANIC SULFUR COMPOUNDS AS INSECT ATTRACTANTS
WITH SPECIAL REFERENCE TO HOST SELECTION

Yoshiharu Matsumoto

Laboratory of Applied Entomology
Faculty of Agriculture
University of Tokyo
Bunkyo-ku, Tokyo

Table of Contents

Table of Contents--(Continued)

CONTROL OF INSECT BEHAVIOR

I. Introduction

The first discussion of the attractiveness of volatile
sulfur compounds for insects in relation to their host selection was published by Verschaffelt in 1910. He reported the
feeding response of Pieris brassicae (L.) to a mustard oil
glucoside, sinigrin. Almost 40 years later, Dethier (1947)
urged further studies of the behavior of insects to sulfur
compounds. Thorsteinson (1953) demonstrated the feeding
stimulation of sinigrin, sinalbin, and glucocheirolin--pre-
cursors of mustard oils--for the diamondback moth larva,
Plutella maculipennis (Curt.).
Illustrations of volatile sulfur compounds acting as insect attractants are found in host selection studies beginning in the latter half of the 1950's. Based on the early
work of Thorsteinson (1953), Gupta and Thorsteinson (1960a,
b) demonstrated the attraction and oviposition stimulation
of allyl isothiocyanate for the diamondback moth. Recently
Traynier (1965, 1967) reported that methyl isothiocyanate
and allyl isothiocyanate enhanced oviposition of the cabbage
maggot, Erioischia brassicae (Bouché). In this paper two
studies are described: the responses of the vegetable weevil, Listroderes obliquus (Klug), to mustard oils (iosthio-
cyanates) (Sugiyama and Matsumoto, 1957, 1959a) and the responses of the onion maggot, Hylemya antiqua (Meig.), to sulfides and mercaptans (Matsumoto and Thorsteinson, 1968a, b).

II. Effect of Mustard Oils on Vegetable Weevil Behavior

The food plants of the vegetable weevil include 178
species or varieties in 34 plant families. For this polyphagous insect, however, little was known about host selection before our studies (Sugiyama and Matsumoto, 1957). It
was generally accepted that specific chemical stimuli are
not required to initiate biting by polyphagous insects, but
that plants are sampled randomly until an acceptable one is
encountered (Dethier, 1953; Wardle, 1929). To test this
hypothesis, we initiated studies of the olfactory responses
of the vegetable weevil to mustard oils.
First a survey and classification of food plant records
of this insect was made (Matsumoto, 1961). As a result of
the classification of over 150 plant species, it was found
that Compositae ranked first in number of species (including
varieties), followed by Cruciferae and Umbelliferae
(Table I). This is not to say that a greater number of

Table I

Food Plant List of the Vegetable Weevil, Listroderes obliquus (Klug)

Family	No.	Family	No.	Family	No.
Compositae	46[a]	Rosaceae	3	Piperaceae	1
Cruciferae	26	Oxalidaceae	3	Violaceae	1
Umbelliferae	14	Oenotheraceae	3	Loganiaceae	1
Caryophyllaceae	10	Geraniaceae	3	Amaranthaceae	1
Leguminosae	8	Convolvulaceae	2	Urticaceae	1
Polygonaceae	8	Plantaginaceae	2	Portulaceae[b]	1
Solanaceae	7	Moraceae	2	Primulaceae[b]	1
Labiatae	6	Malvaceae	2	Boraginaceae[b]	1
Liliaceae	5	Verbenaceae	2	Crassulaceae[b]	1
Chenopodiaceae	5	Polemoniaceae	1	Tropaeolaceae[b]	1
Cucurbitaceae	4	Papaveraceae	1	Caricaceae[b]	1
Scrophulariaceae	4				

(Total: 34 families, 178 species)

[a] Numbers of species (including varieties).
[b] These plants are recorded as food plants only by laboratory experiments.
(Modified from Matsumoto, 1961).

food plant species in certain plant families indicates that the insect prefers (or chooses) those plants among the many plants recorded. However, it is a fact that Cruciferae and Umbelliferae contain plants which are severely damaged by this insect. For example, the vegetable weevil used to be called 'turnip weevil' in the U.S.A., or 'carrot weevil' (translated literally from Japanese) in Japan. At any rate, the cruciferous and also umbelliferous plants are recognized as important food plants of this insect.

   The attractiveness of volatile mustard oils which are regarded as the principal odors characterizing cruciferous plants, and distributed most specifically in Cruciferae, was tested initially (Sugiyama and Matsumoto, 1957, 1959a).

A.   Attraction of Newly-Hatched Larvae

   A modification of the Munakata-Ishii method (1954) for screening the rice stem borer attractants was used as a test method. A small piece of absorbent cotton containing 0.05 ml methanol solution of a definite concentration of mustard oil was placed at the closed end of each of the small glass tubes (5 mm x 3 cm). After the evaporation of methanol at 65°C for 30 min, 4 tubes thus treated were placed crosswise in a petri dish (9 x 2 cm) with the opening 1.5 cm distant from, and facing, the center. The tubes were fixed with cellophane tape to the bottom of the dish. Thirty unfed larvae, which had hatched at 25°C within the previous 24 hr, were transferred with a moistened sable brush to a small disc of filter paper (5 mm dia) moistened with distilled water and placed at the center of the dish. A tightly fitting disc of filter paper (11 cm dia) was placed in the lid of the dish to prevent larval escape. Five replicate dishes were set up in each test with a definite concentration of mustard oil and placed in darkness at 25°C. Five other dishes containing tubes treated with methanol, which was allowed to evaporate, were used as the controls. The number of larvae which entered the tubes was counted after 1 hr. The response percentages were calculated by means of the correction with Abbott's formula.

   Significant effects of allyl isothiocyanate and phenyl isothiocyanate are shown in Fig. 1 indicating that the former exceeds the latter in attractiveness. With a modified test method, the relative attractiveness of 7 other mustard oils was determined (Table II). Generally, attractiveness increased with an increasing number of CH_2 groups

137

Table II

Attractiveness of Various Mustard Oils[a] to Newly-Hatched Larvae of Listroderes obliquus at Optimal Dosages

Mustard oil		Boiling point (°C)	Dosage (mg)	Total entering tubes* Odorous : Control	
Methyl isothiocyanate	CH_3NCS	119	1.0	41	: 2
Ethyl "	C_2H_5NCS	131	1.0	54	: 3
Allyl "	CH_2CHCH_2NCS	151	3×10^{-2}	71	: 0
iso-Butyl "	$(CH_3)_2CHCH_2NCS$	160.5	1×10^{-1}	74	: 2
n-Butyl "	C_4H_9NCS	166	1×10^{-1}	71	: 0
Phenyl "	C_6H_5NCS	220	3×10^{-3}	46	: 3
Benzyl "	$C_6H_5CH_2NCS$	245	1×10^{-4}	94	: 1
β-Phenylethyl "	$C_6H_5CH_2CH_2NCS$	255.5	3×10^{-5}	89	: 2
α-Naphthyl "	$C_{10}H_7NCS$	mp. 58	1×10^{-2}	21	: 2

a Two odorous tubes and 2 control tubes were set crosswise in each dish. The total is the number of larvae which entered the tubes in five dishes after 1 hr at 25°C in the dark.

* L.S.D.(0.05): 10.7 (From Sugiyama and Matsumoto, 1959a).

or boiling point. Throughout these experiments, the response of larvae to each mustard oil at increasing dosages formed a bell-shaped curve; i.e., at the higher dosages the responses were reduced.

Some of these effective mustard oils occur in the food plants of this insect, e.g., allyl isothiocyanate in Brassica cernua Thub., B. juncea Coss., B. nigra Koch., B. oleracea var. botrytis L., B. oleracea var. gemmifera DC., B. oleracea var. gongylodes DC., B. rapa L., Capsella bursa-pastoris Medic. and Wasabia wasabi Makino; benzyl isothiocyanate in Lepidium sativum L. and Tropaeolum majus L.; β-phenylethyl isothiocyanate in Brassica alba Boiss., B. nigra Koch. and B. rapa L.

B. Attraction and Biting Stimulation of Adults

The vegetable weevil adult has a clear thigmotactic behavior. When an adult is released in a container, it always walks along the corner line unless otherwise stimulated (Fig. 2). Using this behavior, a simple assay method was devised. Two watch glasses (6 cm dia) were joined together, making the shape of a convex lens. A small piece of absorbent cotton treated with mustard oil was placed in the center of this container, and the behavioral response of a confined adult was observed for 6 min at 20 - 25° C. The adults had emerged from rearing at 20° C on the day preceding the test and were held in the dark at 20° C for about 24 hr with access to water, but not to food. Between 1 and 2 hr before the test, they were transferred into dry dishes in the light to activate their locomotion. Immediately after release in the container, the adult began continuous circular movements along the contact line of the two watch glasses. However, when the cotton treated with mustard oil was present, the adult exhibits a distinctive behavior pattern in response to the odor. After a short time of circular movements, the adult stops walking for a moment, extending and holding up its antennae in a tense manner. Then it slowly turns its head toward the odor source and walks to it, waving the antennae up and down. As it approaches the cotton, it taps the glass surface many times with the tips of antennae which are bent downward at the pedicels. At the same time the head is lowered, and in some cases mandibular movements (biting) are observable. Finally, the adult reaches the cotton and attacks it, continuing the antennal tapping and biting. Later the antennal tapping proved to elicit a successive vibration on the maxillary palps forming a reflex arc between the tip of the antenna and the maxillary palp, each side being independent of the other (Matsumoto, 1967). In some cases, especially at higher dosages, a unique behavior was observed, i.e., adults began to approach the odor source but returned to the circular movements before reaching it. After repeating this behavior many times, adults finally reached the odor source and attacked it, probably due to adaptation to odor.

Five of 6 mustard oils tested were significantly effective in attracting adults and eliciting a biting response (Table III). n-Butyl isothiocyanate evoked the highest positive response among these mustard oils. Although the adult response to benzyl isothiocyanate was not as high as most of the other mustard oils tested, this compound was the most potent for newly-hatched larvae. These differences in response between adults and larvae may have resulted from the diversity of assays used. Also, there may possibly be some conditioning of adults by the food in the larval stage, although the mustard oils in the larval food, chinese cabbage leaves, are not known.

III. Effect of Sulfides and Mercaptans
on Onion Maggot Behavior

Dethier (1947) suggested that a study of typical Allium-feeders to the odorous sulfur compounds of plant origin would be of interest. Progress had been hindered by lack of knowledge of the chemical composition of onion odors until Carson and Wong (1961) identified hydrogen sulfide, n-propyl mercaptan, methyl disulfide, methyl n-propyl disulfide, n-propyl disulfide, methyl trisulfide, methyl n-propyl trisulfide and n-propyl trisulfide. In addition, ethyl alcohol, n-propyl alcohol, isopropyl alcohol, acetaldehyde, propionaldehyde, n-butyraldehyde, acetone and methyl ethyl ketone were also obtained. Neither allyl sulfide nor allyl n-propyl disulfide, which were reported by earlier workers, was detected.

The onion maggot is largely restricted to the onion. The females always selected onion plants for oviposition when onions were planted among other plant species (Matsumoto and Thorsteinson, 1968a; Workman, 1958), even though larvae can develop on other diets not derived from onions (Friend and Patton, 1956; Workman, 1958). Therefore, the effects of some organic sulfur compounds, alcohols

Table III

Response[a] of Adult Listroderes obliquus to Several Mustard Oils

Mustard oil (mg)	Numbers exposed	Numbers responding			Per cent (A+B)
		Retreat	Approach[b] (A)	Attack[b] (B)	
Allyl isothiocyanate					
3.0	20	0	2 (1)	3 (2)	25
1.0	20	1	10 (0)	1 (1)	55
0.6	20	1	5 (1)	1 (1)	30
iso-Butyl isothiocyanate					
3.0	20	0	2 (0)	0 (0)	10
1.0	20	0	4 (0)	4 (2)	40
0.6	20	0	2 (1)	4 (2)	30
n-Butyl isothiocyanate					
1.0	20	0	3 (0)	5 (5)	40

Table III--(Continued)

Mustard oil (mg)	Numbers exposed	Numbers responding			Per cent (A+B)
		Retreat	Approach[b] (A)	Attack[b] (B)	
n-Butyl isothiocyanate					
0.6	20	0	5 (1)	15 (15)	100
0.3	20	0	8 (0)	2 (1)	50
Phenyl isothiocyanate					
0.1	20	1	0 (0)	0 (0)	0
0.06	20	1	1 (0)	0 (0)	5
0.03	20	2	1 (0)	0 (0)	5
0.01	20	1	0 (0)	0 (0)	0
Benzyl isothiocyanate					
0.006	20	3	4 (0)	0 (0)	20

Table III--(Continued)

Mustard oil (mg)	Numbers exposed	Numbers responding			Per cent (A+B)
		Retreat	Approach[b] (A)	Attack[b] (B)	
Benzyl isothiocyanate					
0.003	20	1	2 (0)	1 (0)	15
0.001	20	0	4 (0)	4 (4)	40
0.0006	20	1	4 (0)	1 (1)	25
β-Phenylethyl isothiocyanate					
0.03	20	2	8 (3)	0 (0)	40
0.01	20	3	7 (4)	6 (5)	65
0.006	20	2	5 (1)	7 (4)	60
0.003	20	1	5 (0)	2 (1)	35

a Retreat: Adults walked toward the odor source, but returned to circular movements before reaching it. Approach: Adults approached to within 1 cm of the odor source. Attack: Adults attacked the odor source exhibiting antennal tapping. b Numbers in parentheses: Numbers of adults exhibiting a biting response. (Abstracted from Matsumoto unpublished data).

143

and carbonyl compounds on the oviposition of adults and on
the orientation of larvae were investigated (Matsumoto and
Thorsteinson, 1968a,b).

A. Attraction and Oviposition Stimulation of Adult Females

Fifty mature female Hylemya antiqua were exposed simultaneously to 2 sand dishes (5 cm dia.) in each cage (22.5 x
22.5 x 15 cm), one containing 30 gm of sand treated with
the test chemical and the other untreated. After 16 - 17
hr , the eggs laid in the sand were counted. The sulfur
compounds were applied as a methanol solution and, after
evaporating the methanol at room temperature, 7 ml of distilled water were added. Carbonyl compounds and alcohols
were applied as water solutions.

The females laid a very highly significantly larger
number of eggs in sand dishes treated with n-propyl mercaptan than in untreated dishes at all concentrations tested
(Table IV). N-propyl disulfide caused increased oviposition
in dishes treated with 0.001 and 0.0005 ml, but decreased
oviposition at the higher dosage of 0.01 ml. Methyl disulfide was not favorable for oviposition. Using the presumed
optimal dosage of each chemical, a comparison was made between n-propyl mercaptan and n-propyl disulfide but the
difference observed was not significant. In the tests of
4 alcohols and 5 carbonyl compounds which are recorded as
constituents of onion odor, only n-propyl alcohol and acetone slightly stimulated the oviposition. In the experiments for simultaneous comparisons of n-propyl disulfide or
n-propyl mercaptan with alcohols or carbonyl compounds at
the presumed optimal dosage, much larger numbers of eggs
were always obtained in dishes treated with n-propyl disulfide or n-propyl mercaptan.

The effect of mixing effective organic sulfur compounds
with either alcohol or acetone on the oviposition response
was tested. Only the mixture of n-propyl disulfide with
n-propyl alcohol elicited a significantly increased response
as compared with the single sulfur compound, while ethyl
alcohol reduced the response to both n-propyl disulfide
and n-propyl mercaptan (Table V). Although acetone alone
was slightly stimulating, in mixture it reduced the response
to n-propyl mercaptan.

The landing and ovipositor probing response to n-propyl disulfide and n-propyl mercaptan were observed using a

Table IV

Mean Numbers[a] of Eggs Laid by _Hylemya antiqua_
in Dishes of Sand Treated with Organic Sulfur Compounds

Chemical (ml)	Treated	Untreated	Difference
n-Propyl disulfide			
0.01	152.3	310.5	-158.2**
0.003	173.7	145.5	28.2
0.001	313.8	100.9	212.9**
0.0005	246.0	97.2	148.8***
0.0001	171.7	131.5	40.2
Methyl disulfide			
0.01	25.3	82.4	-57.1***
0.005	153.6	214.2	-60.6
0.001	153.2	168.4	-15.2
0.0001	70.6	80.8	-10.2

Table IV--(Continued)

Chemical (ml)	Treated	Untreated	Difference
n-Propyl mercaptan			
0.2	350.5	145.7	204.8***
0.1	578.6	199.8	378.8***
0.01	402.6	130.6	272.0***
0.005	446.8	123.5	323.3***
0.001	96.7	16.5	80.2***

a Ten replicates.
** Significant at 1% level.
*** Significant at 0.1% level.
(From Matsumoto and Thorsteinson, 1968a).

Table V

Influence of Adding Alcohols or Acetone to Organic Sulfur Compounds
on Oviposition Response[a] by Hylemya antiqua

Expt. No.	Treatment[b]	Response	Difference
1.	n-Propyl disulfide + Ethyl alcohol	98.5	-137.8*
	n-Propyl disulfide	236.3	
2.	n-Propyl disulfide + n-Propyl alcohol	179.1	74.2**
	n-Propyl disulfide	104.9	
3.	n-Propyl disulfide + Acetone	138.8	14.0
	n-Propyl disulfide	124.8	
4.	n-Propyl mercaptan + Ethyl alcohol	81.5	-175.1*
	n-Propyl mercaptan	256.6	
5.	n-Propyl mercaptan + n-Propyl alcohol	206.9	7.9
	n-Propyl mercaptan	199.0	
6.	n-Propyl mercaptan + Acetone	121.9	-118.6***
	n-Propyl mercaptan	240.5	

[a] Number of replicates: 8. [b] Dosages: n-propyl disulfide=0.001 ml, n-propyl mercaptan= 0.005 ml, ethyl alcohol=0.7 ml, n-propyl alcohol=0.07 ml, acetone=0.7 ml. * Signifi- cant at 5% level. ** Significant at 1% level. *** Significant at 0.1% level. (From Matsumoto and Thorsteinson, 1968a).

copper screen placed on a treated dish. More females landed
on the screen under which these compounds were placed than
on the control screen (Table VI). Moreover, after landing,
the flies which landed on the attractive screen walked in
an excited manner and some of them repeatedly probed the
screen surface with the extended proboscis. Finally, the
ovipositor was inserted through the mesh of the screen.
This was usually repeated sequentially from mesh to mesh although no egg was obtained during the 20 min observation.
The flies on the control screen were much less active and
probing and insertion of the ovipositor rarely occurred.

Similar results were obtained with moistened commercial
onion powder.

B. Attraction of Newly-Hatched Larvae

Chemotactic responses of newly-hatched larvae to organic sulfides and mercaptans were also investigated using
a modification of the method for the vegetable weevil larvae
described earliẻr (Matsumoto and Thorsteinson, 1968b).

Two small glass tubes (5 mm x 3 cm) with a small piece
of absorbent cotton at the closed end were set at right angles on filter paper (7 cm dia) moistened with distilled water and placed in a petri dish (9 x 2 cm). A small amount
(0.2 - 0.5 µl) of test chemical was added to the cotton in
one of the two tubes with a micro-pipette. The moistened
filter paper provided a satisfactory surface for larval
crawling. About 60-90 larvae were placed on a small filter
paper disc (5 mm dia) in the center of the petri dish and
the dish was covered. Usually newly-hatched larvae aggregate in masses so it was not convenient to use a constant
number for each trial. The number of larvae reaching the
mouth of the tubes was recorded for 10 min at 22 - 25°C. At
the end of each observation period the number of larvae that
had entered the tubes was counted.

When the dosage of a chemical was optimal, most of the
larvae quickly crawled straight toward the mouth of the
tube, immediately entered, and crawled into the cotton at the
end of the tube (Fig. 3). Many, however, passed around the
mouth area but some of these eventually returned.

Twenty-seven compounds elicited positive orientation of
the larvae. In addition to those reported in Table VII,
the following chemicals were effective:
Sulfides: Ethyl sulfide, n-Butyl sulfide, iso-Butyl sulfide,
           n-Butyl methyl sulfide, n-Butyl ethyl sulfide,
           iso-Pentyl sulfide.

148

Table VI

Mean Numbers of Hylemya antiqua Females Landing on and Inserting Ovipositor Through Screens Above Dishes Treated with n-Propyl Disulfide or n-Propyl Mercaptan in One Minute Intervals During 20 Minutes

Treatment	Landing on screen above[a]			Ovipositor insertion above[a]		
	Treated	Untreated	Difference	Treated	Untreated	Difference
n-Propyl disulfide (0.005 ml)	2.70 (54)	1.00 (20)	1.70***	2.90 (58)	0.40 (8)	2.50***
	2.15 (43)	0.80 (16)	1.35***	1.35 (37)	0.05 (1)	1.30***
	3.65 (73)	1.75 (35)	1.90***	0.70 (14)	0.30 (6)	0.40
	2.60 (52)	1.60 (32)	1.00*	2.55 (51)	0.20 (4)	2.25***
n-Propyl mercaptan (0.005 ml)	2.95 (59)	1.60 (32)	1.35*	0.35 (7)	0.05 (1)	0.20*
	4.15 (83)	1.65 (33)	2.50***	1.50 (30)	0.00 (0)	1.50***
	4.80 (96)	2.05 (41)	2.75***	1.20 (24)	0.10 (2)	1.10***
	2.15 (43)	0.90 (18)	1.25***	1.10 (22)	0.00 (0)	1.10***

a Numbers in parentheses: Total number of flies observed during 20 min. * Significant at 5% level. *** Significant at 0.1% level. (Abstracted from Matsumoto and Thorsteinson, 1968a).

149

Table VII

Olfactory Response of Newly-hatched Larvae of Hylemya antiqua to Organic Sulfur Compounds

Test Compounds	Dish No. 1	2	3	4	5	6
	Numbers of larvae aggregated at tube mouth and entering tubes[a]					
Sulfides						
Allyl sulfide $(CH_2:CHCH_2)_2S$	69(39)	51(27)	41(27)	53(24)	37(14)	52(28)
Control	3(0)	3(0)	3(0)	6(1)	3(0)	1(1)
n-Propyl sulfide $(C_3H_7)_2S$	57(41)	49(26)	70(41)	65(58)	59(20)	44(31)
Control	3(0)	2(1)	4(2)	7(2)	9(0)	4(3)
Disulfides						
Methyl disulfide $(CH_3S-)_2$	66(25)	75(6)	74(8)	77(9)	79(31)	65(13)
Control	5(0)	4(0)	3(1)	5(0)	7(0)	1(1)
n-Propyl disulfide $(C_3H_7S-)_2$	33(16)	33(16)	41(33)	40(10)	27(14)	35(18)
Control	3(0)	11(0)	8(0)	1(0)	2(0)	2(0)
Mercaptans						
n-Propyl mercaptan C_3H_7SH	55(22)	30(16)	44(6)	42(19)	21(17)	56(27)
Control	11(0)	4(0)	1(0)	8(1)	2(0)	2(0)

Table VII--(Continued)

Test Compounds	Numbers of larvae aggregated at tube mouth and entering tubes[a]					
	Dish No. 1	2	3	4	5	6
Mercaptans--(Continued)						
iso-Propyl mercaptan $(CH_3)_2CHSH$	45(34)	53(35)	37(34)	40(27)	51(48)	53(10)
Control	7(0)	4(0)	4(0)	4(0)	4(0)	6(0)

[a] Numbers in parentheses: Number of larvae entering tubes. (Modified from Matsumoto and Thorsteinson, 1968b).

Disulfides: Ethyl disulfide, n-Butyl disulfide, iso-Butyl
              disulfide, tert-Butyl disulfide, n-Pentyl
              disulfide, iso-Pentyl disulfide.
Mercaptans: n-Butyl mercaptan, n-Pentyl mercaptan, iso-
              Pentyl mercaptan, n-Hexyl mercaptan, n-Heptyl
              mercaptan, Benzyl mercaptan, m-Tolyl mercaptan,
              2-Phenylethyl mercaptan, 2-Hydroxyethyl
              mercaptan.

C. Field Trapping Studies

Field traps were baited with 9 volatile organic sulfur
compounds in the summer of 1964 at Winnipeg, Canada (Matsumoto and Thorsteinson, unpublished data). The trap was
made of a transparent plastic cylinder (13 cm dia x 13 cm
high). A cone (6.5 cm high) with an entrance hole (6 mm
dia) was fitted with plaster of Paris to the bottom of the
trap. The trap was supported about 5 cm from the ground
by a triangular wire frame inserted into the soil. A petri
dish (9 x 2 cm) containing 12 ml of a 10% (v/v) mixture of
each test chemical in odorless paraffin oil was covered
with a plastic screen top and placed beneath the trap.
Traps were placed for 3 days 20 feet apart on fallow land
in a latin square design of 3 chemicals and 1 control in
each trial.

As shown in Table VIII, n-propyl disulfide particularly,
and also n-propyl mercaptan, consistently captured many
gravid female H. antiqua. The reason for the preponderance
of gravid females (up to 90% of the catch) in the captures
was not elucidated.

Methyl disulfide, a constituent of onion odor, was
found to be a highly powerful lure for the black blowfly,
Phormia regina (Meig.) (Table VIII), confirming Scott's
observation (1961). However, this compound was not attractive to the onion maggot. The author was surrounded by the
humming of an exceptionally large number of flies on the
test field as soon as methyl disulfide was transferred into
the dishes beneath the trap. During the first 10 min after
placing the chemical, more than 50 flies were captured in
each trap. A maximum of more than 700 flies was captured
in one day. However, the lure effect of this compound was
lost by the second day, probably due to its high volatility.
More than 95% of the captures were females and numerous egg
masses were deposited inside the trap. However, it has

Table VIII

Field Trapping Tests with Organic Sulfur Compounds in Summer, 1964, at Winnipeg, Canada

Date	Compounds	Number trapped					
		Hylemya antiqua		Phormia regina		Phyllotreta cruciferae	
		♀	♂	♀	♂	♀	♂
June 30 - July 3	n-Propyl mercaptan	27	2	0	0		
	iso-Propyl mercaptan	2	1	0	0		
	n-Propyl disulfide	80	11	0	0		
	Control	6	0	0	0		
July 5 - 8	Allyl sulfide	5	4	0	0		
	Methyl disulfide	2	0	2,455	30		
	n-Propyl disulfide	47	14	33	0		
	Control	3	0	0	0		
July 10 - 13	Ethyl disulfide	1	0	0	0		
	Methyl disulfide	1	1	2,194	25		
	n-Propyl disulfide	62	3	2	0		
	Control	0	0	0	0		

Table VIII--(Continued)

Number trapped

Date	Compounds	Hylemya antiqua		Phormia regina		Phyllotreta cruciferae	
		♀	♂	♀	♂	♀	♂
July 17 - 20	Ethyl mercaptan	0	2	1	0		
	Methyl disulfide	2	0	1,487	10		
	n-Propyl disulfide	60	3	0	0		
	Control	2	0	0	0		
July 23 - 30	n-Propyl mercaptan	13	2	0	0		
	n-Propyl mercaptan + n-Propyl disulfide	75	4	12	0		
	n-Propyl disulfide	65	9	0	0		
	Control	5	0	0	0		
Sept. 3 - 6	Allyl isothiocyanate	0	0			724	0
	n-Butyl isothiocyanate	0	0			23	0
	n-Propyl disulfide	1	0			0	0
	Control	0	0			0	0

(From Matsumoto and Thorsteinson, unpublished data).

154

CONTROL OF INSECT BEHAVIOR

not yet been established that the chemical directly stimulates the oviposition. Other blowfly species, e.g.,
Phaenicia sericata (Meig.), Lucilia illustris (Meig.) and
Callitroga macellaria (Fabricius), were sporadically trapped
with methyl disulfide. Cragg and Thurston (1950) reported
that ethyl mercaptan and methyl disulfide in mixture with
hydrogen sulfide or carbon dioxide formed powerful attractants for females of the blowflies, Lucilia caesar (L.), L.
illustris and P. sericata. However, each sulfur compound
was not very effective alone. Our experiments confirm the
results of Cragg and Thurston (1950), showing the selective
attraction of sulfur compounds for females. The reason for
this differential attraction has not been clarified. A flea
beetle, Phyllotreta cruciferae (Goeze), was attracted to
allyl and n-butyl isothiocyanate. Scott (1961) also reported trapping a flea beetle, but the species was not identified.

IV. Discussion

Volatile organic sulfur compounds of plant origin
function as token stimuli in the host selection behavior of
both a polyphagous and an oligophagous species. The studies
reported here demonstrate that mustard oils, especially n-
butyl isothiocyanate, elicit both attraction and biting
responses from the vegetable weevil. This dual effect of
one chemical also was shown for the onion maggot, i.e.,
attraction and oviposition were stimulated by n-propyl disulfide and n-propyl mercaptan.

The positive olfactory responses of vegetable weevil
and onion maggot larvae to chemicals in their host plants
indicates that the larvae of these two species can perceive
and select their host plants in nature if the eggs are
laid near the plant. Larval host selection has been reported for Papilio ajax (L.) (Dethier, 1941), Chilo suppressalis
Walker, (Munakata et al., 1959) and Bombyx mori (L.)
(Watanabe, 1958; Hamamura et al., 1961).

Many mustard oils attracted vegetable weevil larvae
and adults and induced biting in adults. Subsequently
volatile substances from umbelliferous plants such as
anethole, linalool and limonene, were also found to be
effective for attracting the larvae (Sugiyama and Matsumoto,
1959b), and for attracting the adults and eliciting biting
(Matsumoto, unpublished data). Further screening of chemicals may reveal additional volatile constituents of plants

155

that induce orientation and biting responses from this
insect. Odorous substances are not always effective;
thymol and carvacrol, which are umbelliferous compounds,
proved to be ineffective as larval attractants. However,
p-cymol, which is an analogue of thymol, is effective.
Coumarin, a flavor of sweet clover, was attractive to
adults but it inhibited their feeding (Matsumoto, 1962).
The precision of the range of odors inducing attraction
and feeding stimulation is supported by the fact that the
attractiveness values of the compounds tested are different.

Of the sulfur compounds which were recorded as constituents of onion odor by Carson and Wong (1961), n-propyl
disulfide, n-propyl mercaptan and methyl disulfide were
tested. The first two compounds were effective in attraction and oviposition stimulation of onion maggot adults,
but methyl disulfide was effective only in attraction of
the larvae. However, further tests with other sulfur
compounds which were reported by Carson and Wong (1961)
should be conducted.

The studies reported here have contributed to a better
understanding of the chemical basis of the host selection
behavior of the vegetable weevil and the onion maggot.
It should now be possible to apply this knowledge to the
survey, forecast and control of these important insect
pests.

CONTROL OF INSECT BEHAVIOR

V. References

Carson, J. F., and Wong, F. F. (1961). Agr. Food Chem. 9, 140.

Cragg, J. B., and Thurston, B. A. (1950). Parasitology 40, 187.

Dethier, V. G. (1941). Amer. Naturalist 75, 61.

Dethier, V. G. (1947). "Chemical Insect Attractants and Repellents" p. 52. Blakiston Co., Philadelphia.

Dethier, V. G. (1953). Symposia IXth Intern. Congr. Ent., Amsterdam, 1951, 81.

Friend, W. G., and Patton, R. L. (1956). Can. J. Zool. 34, 152.

Gupta, P. D., and Thorsteinson, A. J. (1960a). Ent. Exp. Appl. 3, 241.

Gupta, P. D., and Thorsteinson, A. J. (1960b). Ent. Exp. Appl. 3, 305.

Hamamura, Y., Naito, K., and Hayashiya, K. (1961). Nature 190, 879.

Matsumoto, Y. (1961). Jap. J. Appl. Ent. Zool. 5, 245.

Matsumoto, Y. (1962). Jap. J. Appl. Ent. Zool. 6, 141.

Matsumoto, Y. (1967). Appl. Ent. Zool. 2, 31.

Matsumoto, Y., and Thorsteinson, A. J. (1968a). Appl. Ent. Zool. 3, 5.

Matsumoto, Y., and Thorsteinson, A. J. (1968b). Appl. Ent. Zool. 3, 107.

Munakata, K., and Ishii, S. (1954). 2·4-D Research 3, 18.

Munakata, K., Saito, T., Ogawa, S., and Ishii, S. (1959). Bull. Agr. Chem. Soc. Japan 23, 64.

Scott, J. A. (1961). Unpublished M.S. Thesis, Univ. Manitoba, Winnipeg.

Sugiyama, S., and Matsumoto, Y. (1957). Nôgaku Kenkyû 45, 5.

Sugiyama, S., and Matsumoto, Y. (1959a). Nôgaku Kenkyû 46, 150.

Sugiyama, S., and Matsumoto, Y. (1959b). Nôgaku Kenkyû 47, 141.

Thorsteinson, A. J. (1953). Can. J. Zool. 31, 52.

Traynier, R. M. M. (1965). Nature 207, 218.

Traynier, R. M. M. (1967). Ent. Exp. Appl. 10, 401.

Verschaffelt, E. (1910). Proc. Acad. Sci. Amsterdam 13, 536.

Wardle, R. A. (1929). "The Problems of Applied Entomology". McGraw-Hill Book Co., Inc., New York.

YOSHIHARU MATSUMOTO

Watanabe, T. (1958). Nature 182, 325.
Workman, R. B. (1958). Unpublished Ph.D. Thesis, State
    College, Univ. Oregon, Corvallis.

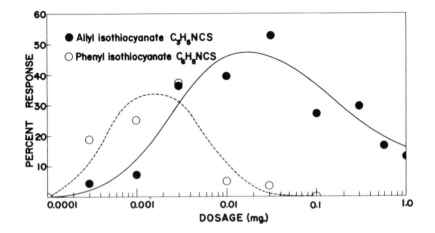

Figure 1.--Olfactory response of newly-hatched larvae of
    Listroderes obliquus to allyl isothiocyanate and
    phenyl isothiocyanate. (From Sugiyama and Matsumoto,
    1957).

158

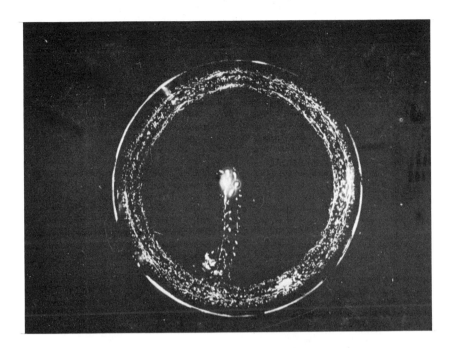

Figure 2.--Footprints of a Listroderes obliquus adult in a
    convex-type container for 10 min. Released at the
    center, the adult moves to the perimeter and begins
    continuous circular movements.

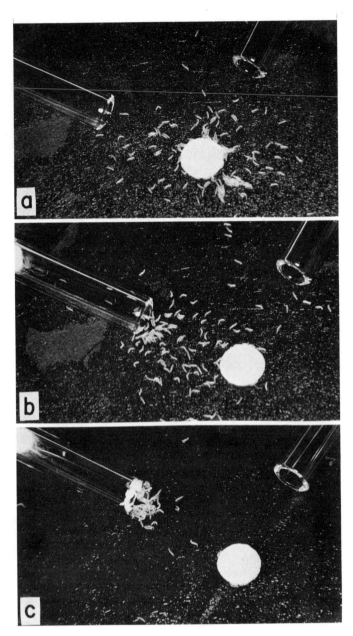

Figure 3.--Orientation of newly-hatched larvae of <u>Hylemia</u> <u>antiqua</u> to n-propyl mercaptan. a) Just after placing larvae on filter paper disc. b) After 30 sec. c) After 60 sec. (From Matsumoto and Thorsteinson, 1968b).

PHEROMONES OF SOCIAL INSECTS

John C. Moser

Southern Forest Experiment Station
Forest Service, U.S. Department of Agriculture
Alexandria, Louisiana

Table of Contents

JOHN C. MOSER

I. Introduction

Insects function like tiny robots programmed to do
specific jobs. Their nervous systems act like biological
computers; they are activated, as if by punch cards, when
their receptors are stimulated. The external receptors
respond to pressure, sound, light, heat, and chemicals. A
chemical or combination of chemicals is called a pheromone
(Karlson and Lüscher, 1959) if it is secreted by an animal
and influences the behavior of other individuals of the
same species. Unlike hormones, pheromones are released into
the environment and transmit their messages between, rather
than within, individuals. This system of communication has
reached its highest development in the social insects.

In this paper, the term social insect as defined by
Richards (1961) is used: "A true social insect may be defined as one in which the female tends or helps to construct
a brood-chamber for an egg (or larva) laid by another
female. This condition is realized only in the ants, bees,
and wasps belonging to the order Hymenoptera, and in the
termites; the latter are unlike the Hymenoptera in that the
males play as big a part in the colony as the females."

II. Primer Pheromones

Pheromones have been divided into two types, primers
and releasers, though the distinction is not always clear-
cut (Wilson, 1963). Primer pheromones trigger a series of
physiological events that change the animal's development
and behavior. Most are transmitted orally, but it is un-
certain whether they act directly on the endocrine system
or first affect chemo-receptors.

A common function of primers in social insects is
caste regulation. One of the queen substances of the honeybee is thus far the only primer that has been synthesized.
Normally, this primer acts like a birth control pill, but
during mating activities it is a sex attractant for drones--
a releaser pheromone. Similar caste inhibitory pheromones
are produced by termites and advanced forms of ants. As a
whole, primer pheromones are poorly understood because of
the small amounts of material available for analysis, and
the lack of suitable bioassays.

162

CONTROL OF INSECT BEHAVIOR

III. Releaser Pheromones

Releaser pheromones cause an immediate response in the
recipient, and seem to be the primary means of communication
in social insects. The best known compounds are those that
trigger mating, alarm, attraction, repulsion, and trail
following. Less understood are those for grooming, food
exchange, gathering and settling of workers, and numerous
other acts of programmed behavior.

A. Alarm

About 20 alarm releasers have been chemically identified from Hymenoptera that have highly developed social
behavior. Hymenoptera of low social development, such as
Ponera and Myrmecina (Formicidae), Polistes (Vespidae), and
Bombus (Apidae), do not possess an alarm mechanism
(Maschwitz, 1966).
Bossert and Wilson (1963) predicted that most alarm
releasers would have between 5 and 10 carbon atoms and
molecular weights between 100 and 200. They envisioned compounds of simple structure and high volatility. The known
alarm substances of termites, however, seem to be of low
volatility and are transmitted by body contact or bumping
(Stuart, 1967). The alarm substance of Nasutitermes is
largely composed of α- and β-pinenes, whereas that for other
genera, such as Coptotermes, does not include pinenes
(Moore, 1965).
Alarm substances of the social Hymenoptera are produced by the mandibular, poison, Dufour's, and anal glands.
Analyses invariably reveal the presence of many compounds
in gland contents. Which is the real pheromone, and what
are the functions of the other compounds?
Recently, three bioassay methods have been proposed to
help answer these questions and to determine behavioral
thresholds. I will call them the molecular diffusion, air
convection, and air-stream methods. In the first, a capillary tube containing a drop of the compound is placed near
an insect, and the threshold levels are calculated from the
elapsed time until the insect responds and the rate of gas
diffusion (Regnier and Wilson, 1968). In the second, known
amounts of pheromone headspace vapor are injected via a
very slowly moving current of air into a large container in
which convection currents further dilute the vapor and carry
it to the ants (Fig. 1) (Moser et al., 1968). The air-

stream method has so far been used only to test sex pheromones (Gaston and Shorey, 1964), but should also work well
for alarm pheromones. An air stream flowing past a paper
disc saturated with the pheromone picks up a known amount
of vapor and then passes the test subject. The number of
molecules/cm^3 required to trigger a given response can be
derived from the airflow rate and the pheromone evaporation
rate.

Of two active compounds found in the mandibular glands
(Fig. 2) of Atta texana (Buckley) (Moser et al., 1968),
4-methyl-3-heptanone (molecular weight 128) causes alarm
with the fewest molecules. Insect response depends upon
the concentration. This ketone acts as an aggregation
pheromone on workers of A. texana at a concentration of
about 33 million molecules/cm^3. A tenfold larger concentration causes alarm, and the liquid placed on trails
repels workers. Responses to far smaller concentrations
of chemicals have been reported. For example, Shorey et al.
(1967) reported that the cabbage looper responded to its
sex attractant at a concentration of 60,000 molecules/cm^3.

Within a species, response may vary widely by caste
and activity. Higher concentrations are required to alarm
workers carrying leaves or detritus than those not so
occupied. Males and queens are not alarmed by 4-methyl-3-
heptanone, although, judging from the intense odor of
crushed heads, their mandibular glands produce more of it
than the workers. Hence, it appears that a caste may emit
a pheromone signal to which it does not respond.

Two or more species may react to a single chemical.
For example, 4-methyl-3-heptanone does not excite
Trachymyremex septentrionalis (McCook), a leaf-cutting ant
closely related to A. texana, but it does alarm Pogonomyrmex
comanche Wheeler, a distant relative. Nests of both species
may be adjacent to those of A. texana (Moser, 1960).

The central nervous system of a species responds to a
specific molecular structure, any deviation from which
decreases sensitivity. For instance, in A. texana the
closer the methyl group was to the fourth carbon atom in
7-carbon ketones, the greater the potency (Table I).
Similarly, the ketone group was most effective on carbon
atom 3; 4-methyl-3-heptanone was 10,000 times more potent
than 4-methyl-2-heptanone. The most effective 7-carbon
compounds without methyl groups were those in which the
ketone groups were on carbon atoms 2 and 3 (Table II).

164

Table I

Response of Atta texana to Ketones with
Different Methyl Group Positions

Ketone	Alarm threshold[a] (molecules/cm^3)
2-methyl-3-heptanone	3.3 X 10^{14}
4-methyl-3-heptanone	3.3 X 10^8
5-methyl-3-heptanone	3.4 X 10^{11}
6-methyl-3-heptanone	3.0 X 10^{13}

[a] Thresholds computed, in part, from boiling-point data
compiled by K. W. Greenlee, Chemical Samples Co.

Table II

Response of Atta texana to Ketones with
Different Ketone Group Positions

Carbonyl	Alarm threshold[a] (molecules/cm^3)
n-heptanal	repellent
2-heptanone	5.3 X 10^{11}
3-heptanone	6.7 X 10^{11}
4-heptanone	7.7 X 10^{14}

[a] Thresholds computed, in part, from boiling-point data
compiled by K. W. Greenlee, Chemical Samples Co.

165

When the ketone and methyl positions were reversed (3-methyl
-4-heptanone), sensitivity was decreased by 10,000 times.
Seven-carbon chains evoked stronger responses than shorter
or longer chains (Table III).

Table III

Response of Atta texana to Ketones with
Varying Number of Carbon Atoms

Ketone	Alarm threshold[a] (molecules/cm^3)
4-methyl-3-pentanone[b]	2.4 X 10^{14}
4-methyl-3-hexanone	1.1 X 10^{12}
4-methyl-3-heptanone	3.3 X 10^8
3-pentanone	4.7 X 10^{15}
3-hexanone	1.7 X 10^{14}
3-heptanone	6.7 X 10^{11}
3-octanone	2.6 X 10^{11}
3-nonanone	8.7 X 10^{11}
3-decanone	2.8 X 10^{12}

[a] Thresholds computed, in part, from boiling-point data
compiled by K. W. Greenlee, Chemical Samples Co.
[b] Also 2-methyl-3-pentanone.

Of 35 ketones tested, 2-heptanone, 3-heptanone, 4-methyl-
3-hexanone, 3-octanone, and 3-nonanone were highly active,
though 1000 times less so than 4-methyl-3-heptanone. When
the concentration was increased sufficiently, all 35 ketones
produced alarm.

To comprehend the chemical concentrations at which ants were responding, I compared their detection threshold with my own. We had about equal ability to detect Chanel No. 5. While I could sense approximately equal concentrations of the perfume and the three ketones, the detection thresholds for the ants varied widely (Table IV).

Table IV

Detection Thresholds of Ketones by
Atta texana and Man

Ketone	Detection threshold[a] (molecules/cm^3)	
	A. texana	Moser
4-methyl-3-heptanone	3.3×10^7	1.0×10^{14}
2-heptanone	5.3×10^{10}	1.6×10^{14}
6-methyl-2-heptanone	2.8×10^{13}	8.4×10^{13}

a   Thresholds computed, in part, from boiling-point data compiled by K. W. Greenlee, Chemical Samples Co.

Their sensitivity to 4-methyl-3-heptanone was almost 10^7 times as great as mine.

The alcohol 4-methyl-3-heptanol produced alarm in A. texana, but at a concentration 100,000 times higher than the related ketone. The aldehydes n-heptanal and n-hexanal were repellents to A. texana, as was citral, which is abundant in the mandibular glands of the South American species, A. sexdens L. Citral is an aggregation pheromone in honeybees (Shearer and Boch, 1966; Butler, 1967), and thus, an example of a pheromone that triggers different behavior in different animals.

The aggregation releaser cis-3-hexenol (Verron, 1963) is formed during the breakdown of wood by symbionts of Kallotermes termites. In these species, wood is digested primarily by larvae, which then trophallactically feed

other castes that do not digest wood. These larvae are
highly attractive to the other castes, which possess little
or no such attractiveness.

Both A. texana and honeybees manufacture 2-heptanone
in their mandibular glands. This ketone alarms both species,
but at far higher concentrations than other chemicals
that these insects produce (Moser et al., 1968). It is
known to be the primary alarm substance for Iridomyrmex
pruinosus (Roger) and Conomyrma pyramicus (Roger) (Blum
and Warter, 1966; Blum et al., 1966).

The same pheromone may be produced in different glands
in different species. Thus, in I. pruinosus and C.
pyramicus, 2-heptanone is produced in the anal gland, while
in A. texana and the honeybee it is produced in the mandibular glands.

B. Sex

Sex releasers appear to be located in the mandibular
glands of the higher Hymenoptera. Trans-9-hydroxydec-2-
enoic acid, one of the queen substances of the honeybee,
is located in the mandibular glands of the female, and
attracts males (Gary, 1962). Hölldobler and Maschwitz
(1965) reported that male carpenter ants, Camponotus
herculaneus L., secrete a substance with a resinous odor
from their mandibular glands into the air above the nest.
This substance draws out the females and causes them to
swarm. In sawflies, which are lower Hymenoptera, sex
releasers are present in the abdomen of females (Coppel
et al., 1960). No termite sex releasers have been reported.

C. Trail Following

The only trail-marking pheromone of any social insect
that has been identified and synthesized is that of the
southern subterranean termite, Reticulitermes virginicus
(Banks) (Matsumura et al., 1968). The compound is an
olefinic primary alcohol, n-cis-3, cis-6, trans-8-
dodecatrien-1-ol. Less than 0.1 picogram streaked across a
10 cm glass plate releases trail following. The compound
also exists in wood rotted by Lenzites trabea Pers. ex Fr.
It is not yet known if termites obtain the chemical by
ingesting the fungus or fungus-infected wood, or if they
must synthesize it.

Moore (1966) isolated and characterized the trail-marking substance of Nasutitermes corniger (Motschulsky) as an unsaturated diterpenoid hydrocarbon $C_{20}H_{32}$. Stuart (1963) showed that the trail substance from the same termite was volatile.

Certain artifacts release trail following in termites. Watanabe and Casida (1963) found that R. flavipes (Koller) followed numerous synthetic chemicals, and Becker (1966) showed that many species readily followed ball-point pen inks containing certain glycols (Fig. 3).

Primitive termites lay trails in response to alarm over such events as a breach of the nest, and nymphs, which rebuild damaged areas, follow these trails (Stuart, 1967). The trails of advanced forms of termites are made in direct response to food sources or other stimuli (Grassé and Noirot, 1951). Trail pheromones of termites are in the sternal gland (Fig. 4), a diffuse structure beneath one or more abdominal sternites (Noirot and Noirot-Timothée, 1965). In Nasutitermes, as little as 10^{-8} gm of the natural product will mark 10 miles of a trail that persists from 1 to several days. However, termites placed near trails of highly concentrated chemical wander at random as if no pheromone were present (Moore, 1965).

Trail pheromones of termites seem to be nonspecific, at least within genera. Stuart (1963) states that several species of Nasutitermes follow a single substance. Smythe et al. (1967) showed that the secretion from sternal glands of R. flavipes and R. virginicus is followed by members of both species as well as by R. hesperus Banks, but not by Zootermopsis angusticollis (Hagen).

Walsh et al. (1965) purified the trail substance from Dufour's gland of the fire ant. Cavill et al. (1967) isolated α-farnesene from Dufour's gland in Aphaenogaster longiceps (F.Sm.), but did not attribute a function to it. Moser and Silverstein (1967) showed that the substance from the poison gland of A. texana (Fig. 5) had both volatile and nonvolatile components that partitioned into the organic phase of a methylene chloride-water system. They concluded that the components are not proteinaceous. The volatile substance lasts about 30 min, and the nonvolatile component about 6 days in air, but under a vacuum of 0.4 mm mercury, the nonvolatile material appears to remain stable indefinitely. The substance for fire ants seems to possess only a volatile fraction, since its trail lasts about 100 sec (Wilson, 1965a).

Trail substances for all ants so far analyzed originate in the gaster except for Crematogaster in which they
are stored in the hind tibia and released by the lower foot
(Leuthold, 1968). In gasters they have been located in the
hind gut of ponerines, dorylines, and formicines, in
Pavan's gland of dolichoderines, and in the poison gland
and Dufour's gland of myrmecines.

In the field, most ant trails are species-specific,
i.e., individuals of one species will not share trails
with those of another. Indeed, further work may show that
many are nest-specific within a species. Among the excep-
tions, the formicine, Camponotus beebei Wheeler, utilizes
arboreal odor trails of a dolichoderine, Azteca chartifex
Forel (Wilson, 1965b). Rettenmeyer (1963a) records a case
of the army ant Eciton hamatum (Fab.) sharing trails with
workers of Atta sp.

Some ant inquilines share trails with great precision
(Akre and Rettenmeyer, 1968; Moser, 1964; Rettenmeyer, 1962,
1963b; Watkins et al., 1967). These followers of ant
trails include silverfish, cockroaches, millipedes, and
blind snakes.

Species that apparently refuse to share trails in the
field often do so readily in the laboratory when extracts
of glands are deposited to form an artificial trail. For
example, most species of the fungus-culturing tribe Attini
follow poison-gland extracts of each other in addition to
that of Daceton armigerum (Lat.), which belongs to another
tribe (Blum and Portocarrero, 1966). Oddly, D. armigerum
does not lay trails, nor does it follow extracts of its
own poison gland. Extracts of the poison gland of
Cardiocondyla nuda minutor Forel, another myrmecine that
does not lay trails but communicates by tandem running
(Wilson, 1959), release trail following in Monomorium
minimum (Buckley) (Blum, 1966). Watkins (1964) showed
that several species of Neivamyrmex share trails and
follow such artifacts as distilled water and 70% ethyl
alcohol. It appears, then, that in addition to the basic
trail substance(s) of each species, a yet-to-be-located
species- or nest-specific chemical may be present after
the trail is laid naturally.

Some Camponotus, and perhaps other ants with large
eyes, use trails in the dark, but navigate by vision when
light is available. Only individual foragers of C.
herculanus are seen in Louisiana by day, but at night I

have seen heavy trail activity rivaling that of A. texana. Members of a colony of C. rasilis (Creighton) in a door of my house make clearly defined trails in dark rooms, but forage individually during the day and in lighted rooms at night.

IV. Conclusions

Insects appear to have pheromone receptors that are triggered easily by a specific molecular structure, less easily by closely related structures, and not at all by most molecules. From these receptors, an electrical impulse is transmitted to the central nervous system to release a particular behavior pattern.

The reception mechanism in insects may not be much different from the mechanism for smell in vertebrates. Amoore (1967) and Amoore et al. (1967) suggest that sensations of smell are determined by reactions to primary odors which are analogous to the primary colors in vision. They further suggest that humans differentiate odors from the molecular shapes of the compounds. Wright (1963) advanced the theory that odor is dependent upon molecular vibration, which suggests that radiation from pheromones rather than the molecules themselves triggers the receptors. Callahan (1968) suggests that this theory may be proved by placing pheromone vapors in masers. Thus, we may some day control insects with radiation instead of chemicals. Although the answers to such basic questions are unknown, the future of pheromones for insect control seems already assured.

V. References

Akre, R. D., and Rettenmeyer, C. W. (1968). J. Kansas
    Entomol. Soc. 41, 166.
Amoore, J. E. (1967). Nature 214, 1095.
Amoore, J. E., Palmieri, G., and Wanke, E. (1967). Nature
    216, 1084.
Becker, G. (1966). Z. Angew. Zool. 53, 495.
Blum, M. S. (1966). Proc. Roy. Entomol. Soc. London (Ser.
    A) Gen. Entomol. 41, 150.
Blum, M. S., and Portocarrero, C. A. (1966). Psyche 73,
    150.
Blum, M. S., and Warter, S. L. (1966). Ann. Entomol. Soc.
    Am. 59, 774.
Blum, M. S., Warter, S. L., and Traynham, J. G. (1966).
    J. Insect Physiol. 12, 419.
Bossert, W. H., and Wilson, E. O. (1963). J. Theoret.
    Biol. 5, 443.
Butler, C. G. (1967). Cambridge Philosophical Soc. Biol.
    Rev. 42, 42.
Callahan, P. S. (1968). Personal communication.
Cavill, G. W. K., Williams, P. J., and Whitfield, F. B.
    (1967). Tetrahedron Letters 23, 2201.
Coppel, H. C., Casida, J. E., and Dauterman, W. C. (1960).
    Ann. Entomol. Soc. Am. 53, 510.
Gary, N. E. (1962). Science 136, 773.
Gaston, L. K., and Shorey, H. H. (1964). Ann. Entomol. Soc.
    Am. 57, 779.
Grassé, P., and Noirot, C. (1951). Annee Psychole 50, 273.
Hölldobler, B., and Maschwitz, U. (1965). Z. Vergleich.
    Physiol. 50, 551.
Karlson, P., and Lüscher, M. (1959). Nature 183, 55.
Leuthold, R. H. (1968). Psyche 75, 233.
Maschwitz, U. W. (1966). In "Vitamins and Hormones"
    (Editors, Harris, R. S., Wool, I. G., and Loraine,
    J. A.) 267. Academic Press, New York and London.
Matsumura, F., Coppel, H. C., and Tai, A. (1968). Nature
    219, 963.
Moore, B. P. (1965). Australian J. Sci. 28, 243.
Moore, B. P. (1966). Nature 211, 746.
Moser, J. C. (1960). Forests and People 10, 30.
Moser, J. C. (1964). Science 143, 1048.
Moser, J. C., and Silverstein, R. M. (1967). Nature 215,
    206.

Moser, J. C., Brownlee, R. C., and Silverstein, R. M. (1968). J. Insect Physiol. 14, 529.
Noirot, C., and Noirot-Timothée, C. (1965). Insectes Sociaux 12, 265.
Regnier, F. E., and Wilson, E. O. (1968). J. Insect Physiol. 14, 955.
Rettenmeyer, C. W. (1962). J. Kansas Entomol. Soc. 35, 377.
Rettenmeyer, C. W. (1963a). Univ. Kansas Sci. Bull. 44, 281.
Rettenmeyer, C. W. (1963b). Ann. Entomol. Soc. Am. 56, 170.
Richards, O. W. (1961). "The Social Insects". (Harper Torchbook TB/542). Harper and Row, New York.
Shearer, D. A., and Boch, R. (1966). J. Insect Physiol. 12, 1513.
Shorey, H. H., Gaston, L. K., and Saario, C. A. (1967). J. Econ. Entomol. 60, 1541.
Smythe, R. V., and Coppel, H. C. (1966). Ann. Entomol. Soc. Am. 59, 1008.
Smythe, R. V., Coppel, H. C., Lipton, S. H., and Strong, F. M. (1967). J. Econ. Entomol. 60, 228.
Stuart, A. M. (1963). Physiol. Zool. 36, 69.
Stuart, A. M. (1967). Science 156, 1123.
Verron, H. (1963). Insectes Sociaux 10, 167.
Walsh, C. T., Law, J. H., and Wilson, E. O. (1965). Nature 207, 320.
Watanabe, T., and Casida, J. E. (1963). J. Econ. Entomol. 56, 300.
Watkins, J. F., II. (1964). J. Kansas Entomol. Soc. 37, 22.
Watkins, J. F., II, Gehlbach, F. R., and Baldridge, R. S. (1967). Southwestern Nat. 12, 455.
Wilson, E. O. (1959). Psyche 66, 29.
Wilson, E. O. (1963). Sci. Am. 208, 106.
Wilson, E. O. (1965a). Science 149, 1064.
Wilson, E. O. (1965b). Psyche 72, 2.
Wright, R. H. (1963). Nature 198, 455.

Figure 1.--Air-convection bioassay of alarm pheromone of Atta texana. From fungus gardens (left and right), ants travel through plastic tubes into the test chamber (center). Pheromones are thus tested under conditions that are as normal as possible.

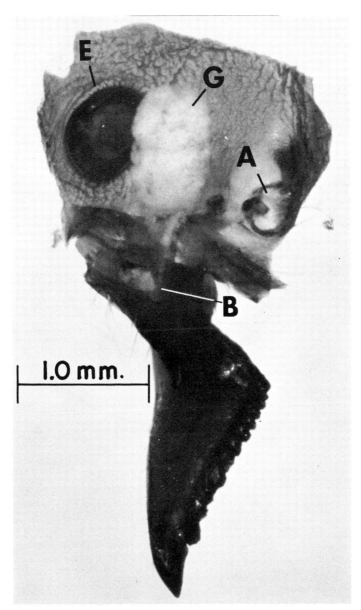

Figure 2.--Interior of the head of Atta texana queen
     dissected to reveal mandibular gland (G) between eye
     (E) and antennal socket (A). Duct leads to mandible
     base (B) where pheromones are released into air.

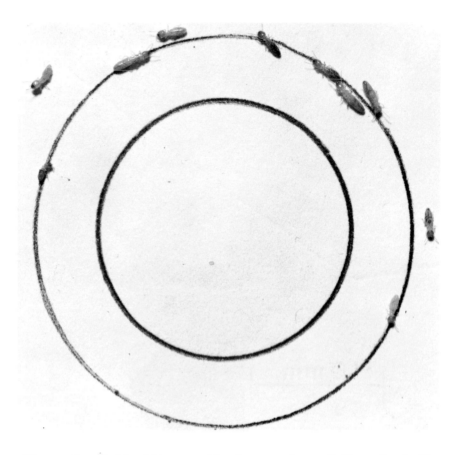

Figure 3.--<u>Reticulitermes</u> <u>flavipes</u> workers follow the ball-
point pen ink of the outer circle but do not respond
to the different ink of the inner circle. (Photo
courtesy of Dr. R. V. Smythe).

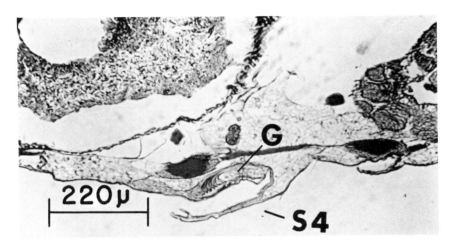

Figure 4.--Position of sternal gland (G) of Reticulitermes flavipes in relation to the elongated fourth abdominal sternite (S4). (Photo courtesy of Dr. H. C. Coppel) (See Smythe and Coppel, 1966).

177

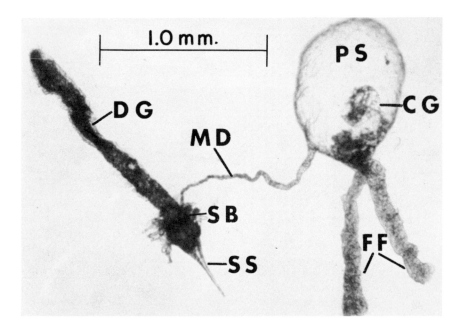

Figure 5.--Poison apparatus of Atta texana worker. (CG) convoluted gland, (DG) Dufour's gland, (FF) free filaments, (MD) main duct, (PS) poison sac, (SB) sting bulb, and (SS) sting shaft.

INSECT ANTIFEEDANTS IN PLANTS

Katsura Munakata

Laboratory of Pesticides Chemistry
Faculty of Agriculture
Nagoya University
Nagoya, Japan

Table of Contents

KATSURA MUNAKATA

I. Introduction

The feeding behavior of insects can be divided into
four steps: (a) host plant recognition and orientation,
(b) initiation of feeding, (c) maintenance of feeding, and
(d) cessation of feeding. An antifeedant is concerned with
steps (b) and (c). The term antifeedant is defined as a
chemical that inhibits feeding but does not kill the insect
directly, i.e., it remains near the treated leaves and dies
through starvation. Gustatory repellent, feeding deterrent
and rejectant are synonymous with antifeedant. Studies of
the natural phenomenon of antifeeding (no consumption of
plant materials) could reveal the presence of new antifeedant compounds, and provide correlations between chemical
structures and antifeeding activities. Antifeedants could
be of great value in protecting crops from noxious insects.

Thorsteinson (1960) reviewed evidence indicating that
host selection in phytophagous insects is governed by the
presence or absence of attractants and repellents in plants.
For example, Buhr et al. (1958) reported that some of the
alkaloid glycosides in plants of the Solanaceae acted as
repellents to the larvae of the Colorado potato beetle,
Leptinotarsa decemlineata Say. Lichtenstein et al. (1962)
and Lichtenstein and Casida (1963) isolated 2-phenylethyl-
isothiocyanate from edible parts of turnip (Brassica rapa
L.) and 5-allyl-1-methoxy-2,3-methylenedioxybenzene or
myristicin from edible parts of parsnip (Pastinaca sativa
L.) as the naturally occurring antifeedants. Moreover, one
of the resistant factors in corn plants to the European
corn borer, Pyrausta nubilalis (Hübner), was identified as
6-methoxybenzoxazolinone by Smissman et al. (1957). Asenjo
et al. (1959) separated a waxy fraction from the extracts
of woods of West Indian mahohany, which showed high termite
repellency. Rudman and Gay (1961) noted that 2-methyl-, 2-
hydroxymethyl-, and 2-formylanthraquinones present in the
extracts of teak heart-wood were all effective in inhibiting
termite activity. The alkaloidal glycosides, such as lepline
II and III, demmissine and tomatine, inhibited feeding of
tomato beetles (Strarchow and Loaw, 1961).

In this paper, antifeedant studies in the author's
laboratory are discussed.

II. Screening Method

In a preliminary experiment, 2 leaf-disks, 16 mm dia,

180

were punched out with a cork-borer from the leaves of food plants. One disk was immersed in an acetone solution of the test sample for 2 min, and the other, the control, in pure acetone. After air drying, these disks were placed symmetrically in a polyethylene dish (100 mm dia x 45 mm deep), and 10 test insect larvae were introduced into the dish. About half the area of the control disk was eaten usually within 2 hr, and the consumed areas of the disks were measured by Dethier's (1947) method. The consumed area of the sample disk expressed as a percentage of the consumed area of the control disks showed the feeding inhibitory activities of the samples. This "leaf-disk test" was repeated twice.

If the feeding rate of larvae is too slow to consume the disks within several hours, the disks dry up and become unsuitable for testing. Then, half leaves, cut along the main vein, were used in place of disks and the test results were checked after 12 hr. When consumption of the sample disk was less than 20% of the consumption of the control disks, the samples were considered to have feeding inhibitory activities.

III. Antifeedants From Plants

A. Antifeedants from. Cocculus trilobus DC.

Cocculus trilobus DC., which is well known as the host plant of Japanese fruit-piercing moths, Oraesia excavata Butler and O. emarginata Fabricius ("Akaeguriba" and "Himeeguriba" in Japanese), is not attacked by any other insects in nature. Therefore, it was assumed that C. trilobus contained toxins or feeding inhibitors against other insects.

Two alkaloids were isolated as crystalline forms from the fresh leaves of this plant. An insecticidal alkaloid was named cocculolidine (II), and an antifeedant alkaloid was identified as isoboldine (I) by the comparison of its physical properties with those of an authentic sample (Kato et al., 1969). The authors proposed structure II for cocculolidine (Wada and Munakata, 1966, 1967; Wada et al., 1967, 1968).

To determine the threshold concentration of isoboldine for feeding inhibition, serial acetone dilutions of pure isoboldine were applied to leaves of Euonymus japonicus Thunberg, and the treated leaves were submitted to the

181

(I) Isoboldine

(II) Cocculolidine

leaf-disk test with Abraxas miranda Butler. The feeding ratios were near zero at concentrations of 200 ppm or greater. At concentrations of 100 and 10 ppm, the feeding ratios were 40 - 59% and 48 - 126%, respectively (Table I). The leaf-disk test with Prodenia litura F. was

Table I

The Feeding Inhibitory Activity of Isoboldine

Insect species	Concentration (ppm)	Per cent consumed		Feeding ratio[a] (B)/(A) X 100
		Control (A)	Treatment (B)	
A. miranda	1000	15	0	0
		35	4	11.4
	500	28	3	10.7
		42	0	0
	200	23	0	0
		22	0	0
	100	39	23	59.0
		20	8	40.0
	10	23	11	47.8
		27	34	126.0

CONTROL OF INSECT BEHAVIOR

Table I--(Continued)

Insect species	Concentration (ppm)	Per cent consumed Control (A)	Treatment (B)	Feeding ratio[a] (B)/(A) X 100
P. litura	200	33	2	6.1
		30	0	0
	100	37	10	26.4
		33	11	33.4
O. excavata	1000	38	38	100.0
		29	29	100.0
	500	65	35	54.0
		66	66	100.0

[a] Strong feeding inhibitory activity, 0-20%; slight, 20-50%; none, 50% and greater.
(Modified from Wada and Munakata, 1968).

conducted with 100 and 200 ppm acetone solutions of pure isoboldine. Leaves of sweet potato, one of the many host plants of this insect, were used as feeding disks. This alkaloid also showed feeding inhibitory activity at 200 ppm, (Table I). The leaf-disk test with O. excavata attacking leaves of C. trilobus was conducted with 500 and 1000 ppm acetone solutions of pure isoboldine, but isoboldine showed no activity.

B.   Antifeedants in Clerodendron tricotomum Thumb.

It has been observed in Japan that the soft leaves of C. tricotomum ("Kusagi", meaning "odorous tree") are not

attacked by insects. Compounds from this plant may prevent
insect attack on rice plants. Feeding tests with P. litura
confirmed that the leaves contained some antifeedants, and
2 colorless antifeedant compounds were isolated and named
clerodendrin A and B respectively: clerodendrin A, mp
163.5-165°, $[\alpha]_D$ +7.4°, (c, 3.3 $CHCl_3$), $C_{31}H_{42}O_{12}$, M^+
606, λmax 203 mu (MeOH in N_2) log ε 4.0; clerodendrin B,
mp 227-229°, $[\alpha]_D$ -82.9°, (c, 0.7 $CHCl_3$), $C_{31}H_{44}O_{12}$, M^+ 608,
λmax 206.5 mu (MeOH in N_2) log ε 3.6. By degradation re-
actions and from mass and n.m.r. spectra, the partial structure of clerodendrin A (III) was identified as follows:

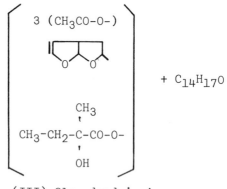

$$3 \ (CH_3CO-O-)$$

$$+ \ C_{14}H_{17}O$$

$$\begin{array}{c} CH_3 \\ | \\ CH_3-CH_2-C-CO-O- \\ | \\ OH \end{array}$$

(III) Clerodendrin A

Clerodendrin B has two more protons than clerodendrin A. The
structures are now being investigated by X-ray analysis. The
minimum concentrations showing antifeeding activity of
clerodendrin A and B for P. litura are under 300 and 200
ppm in acetone solutions, respectively.

C. Antifeedants of Parabenzoin trilobum Nakai. (=Lindera
    triloba Blume)

    Leaves of P. trilobum ("Shiromoji" in Japanese) are not
attacked by the larvae of P. litura. The crude extract also
showed antifeeding activity in the leaf-disk test. The
active antifeedant material was benzene extractable, neutral,
and eluted by 30% ether in n-hexane from a silica gel
column. Two active compounds were isolated from the extract
and named shiromodiol-diacetate (IV) and shiromodiol
monoacetate (V) (Wada et al., 1968). Shiromodiol-diacetate
(IV), $C_{19}H_{30}O_5$, mp 112°C, $[\alpha]_D^{25}$ -61.9°, (c, 1.06 $CHCl_3$),
ν_{max}^{KBr} 1735 and 1240 cm^{-1}. On alkaline hydrolysis, (IV) gave

184

shiromodiol (VI), $C_{15}H_{26}O_3$, mp 89°, which has methyl, iso-propyl and 2 secondary alcohol groups. The reaction pro-ducts of VI on hydrogenation, acetylation, oxidations and dehydrogenation, and their spectral data, suggested a ses-quiterpene having the structure VI.

Shiromodiol-monoacetate (V), $C_{17}H_{28}O_4$, mp 80°, $[\alpha]_D^{25}$ -44.8°, (c, 0.34 $CHCl_3$), ν_{max}^{KBr} 3460, 1700 and 1250 cm^{-1}, was assigned structure V; on acetylation it gave the diacetate (IV) and on oxidation an α-epoxy ketone, mp 144°C, $C_{17}H_{26}O_4$ ν_{max}^{KBr} 1730, 1685 and 1245 cm^{-1}.

The threshold concentration of IV was around 0.125% and 0.033% acetone solution when tested with P. litura and A. miranda, respectively.

IV. Shiromodiol-diacetate. $R_1 = R_2 = Ac$

V. Shiromodiol-monoacetate. $R_1 = H$; $R_2 = Ac$

VI. Shiromodiol. $R_1 = R_2 = H$

IV. Conclusions

We have isolated and identified 5 antifeedant compounds from 3 plant leaves. In nature, these compounds seem to play a role as resistant factors against insect attack. The following methodology is suggested for antifeedant studies:

(1) development of methods of rearing test insect larvae throughout the year;

(2) survey of resistant plants;

(3) preparation of leaf-disk test for monitoring the isola-tion experiments;

(4) preliminary extraction of leaves with several solvents monitored by leaf-disk test. (Approximately 10 g of leaves would usually be used);

(5) collection of large amounts of leaves having antifeed-ing activity. (Up to 10 kg have been used);

(6) isolation experiments monitored by the leaf-disk test;
(7) identification of active compounds, and chemical syn-
    thesis of them and their related compounds;
(8) field test of synthesized compounds.

Antifeedants kill the insects indirectly through
starvation. Antifeedants are not harmful to parasites,
predators or pollinators. If the crops are sprayed by
efficient antifeedants, the insects may turn from crops to
weeds.

V. References

Asenjo, C. F., Marin, L. A., Torres, W., and Campillo, A. (1959). Chem. Abstr. 53, 22707.

Buhr, H., Toball, R., and Schreiber, K. (1958). Ent. Exp. Appl. 1, 209.

Dethier, V. G. (1947). In "Chemical Insect Attractants and Repellents" p. 210. Blakiston Co., Philadelphia.

Kato, N., Shibayama, S., Takahashi, M., and Munakata, K. (1969). Chemical Communications, London. (In press).

Lichtenstein, E. P., Strong, F. M., and Morgan, D. G. (1962). J. Agr. Food Chem. 10, 30.

Lichtenstein, E. P., and Casida, J. E. (1963). J. Agr. Food Chem. 11, 410.

Rudman, R., and Gay, F. J. (1961). Hotzforschung 15, 117.

Smissman, E. E., LaPidus, J. B., and Beck, S. D. (1957). J. Amer. Chem. Soc. 79, 4697.

Strarchow, B., and Loaw, I. (1961). Ent. Exp. Appl. 4, 133.

Thorsteinson, A. J. (1960). Ann. Rev. Entomol. 5, 193.

Wada, K., and Munakata, K. (1966). Tetrahedron Letters No. 42, 5179.

Wada, K., and Munakata, K. (1967). Agr. Biol. Chem. 31, 336.

Wada, K., Marumo, S., and Munakata, K. (1967). Agr. Biol. Chem. 31, 452.

Wada, K., and Munakata, K. (1968). J. Agr. Food Chem. 16, 471.

Wada, K., Enomoto, Y., Matsui, K., and Munakata, K. (1968) Tetrahedron Letters No. 45, 4673.

1,3-DIOLEIN, A HOUSE FLY ATTRACTANT
IN THE MUSHROOM, Amanita muscaria (L.) Fr.

Toshio Muto and Ryozo Sugawara

Laboratory of Pesticide Chemistry
Faculty of Agriculture
Tokyo University of Education
Tokyo, Japan

Table of Contents

Table of Contents--(Continued)

CONTROL OF INSECT BEHAVIOR

I. Introduction

It has long been known that house flies gather on the
fruiting bodies of some mushrooms belonging to the families
Tricholomataceae and Amanitaceae and die. Takemoto and
Nakajima (1964) successfully isolated the toxic components
in these mushrooms, elucidated their chemical structures as
novel acidic amino acids and proposed the names of tricho-
lomic acid and ibotenic acid.

However, we have been more interested in the phenomemon
of fly attraction to the mushroom than in the presence of
toxins. The attractive components have been extracted from
the fruiting body, and one was isolated as colorless crystals (Muto and Sugawara, 1965). The substance was tentatively identified as 1,3-diolein. Its identity was confirmed by comparison of its properties and its ozonolysis and
hydrogenation products with those of an authentic sample
(Daubert and Lutton, 1947).

Several related compounds were prepared and their
activity compared. 1-Monoölein and 2-monoölein were at
least 10 times more attractive than 1,3-diolein. 3-Chloro-
and 3-methoxy-1,2-propanediol 1-monoöleates and 1,2-pro-
panediol 1-monoöleate exhibited nearly the same activity as
1,3-diolein. Glycerol, oleic acid, methyl oleate, triolein,
ethylene glycol dioleate and 1,3-distearin had no significant activity.

From these results, we considered that at least one
esterified residue of an unsaturated, long chain fatty acid
together with at least one hydroxyl group, preferably primary, should be necessary for compounds to exhibit attractancy. We therefore prepared a series of α,ω-glycol mono-
öleates with a general formula, $CH_3(CH_2)_7CH=CH(CH_2)_7COO-$
$(CH_2)_nOH$ (n=2,3,4,5,6,8 and 10), from the appropriate glycols and oleoyl chloride. The activity of the esters with
n = 2 through 6 were found to be at about the same level as
1-monoölein, and a significant drop in activity was noted
for the esters with n = 8 and 10.

In a series of saturated fatty acid esters of ethylene
glycol, the monolaurate showed the highest activity (Muto
et al., 1968).

II. Source of Attractant

We selected Amanita muscaria (L.) Fr. as the material

191

for extraction. This mushroom is native in the northern
mountains of Japan and is characterized by the pink-colored
cap, 10-15 cm dia, spotted with white warts; the stalk may
reach 15 cm in length. The fruiting bodies of the mushroom
were harvested from the end of September through early
October. They were crushed immediately after harvest, filtered through cloth, and the filtrate was adjusted to pH
7-8. The solution was covered with toluene and held at
about 0°C.

III. Assay

A room 325 cm wide, 294 cm deep and 245 cm high, controlled at 26°C and 60% RH, was equipped with a shelf along
the wall 30 cm wide at a height of 62 cm from the floor.
Three 20-watt fluorescent lights were fixed at a height of
53 cm from the shelf. Five hundred to 1,000 house flies,
about 4 days after emergence, were released in the room 30
min before the experiment. Petri dishes, 2 cm deep and 9 cm
dia, containing 20-30 ml of aqueous test solutions, were
placed on the shelf at intervals of about 30 cm alternating
with dishes containing the same volume of water. The petri
dishes were left in the room for a suitable period of time,
usually overnight (16 hr) under illumination. Powdered
milk and a water supply with wicks were placed in the center
of the room. The attractiveness of the samples was based
on the number of flies in the petri dishes at the end of
the test period.

IV. Fractionation

Filtered juice (3000 ml) was adjusted to pH 7 and
filtered through a layer of Celite. The dark brown filtrate
was concentrated to 500 ml under reduced pressure at 60-70°C
while the pH of the solution in the flask was intermittently
adjusted to 7. This procedure was repeated twice more until
the distillate exhibited no detectable attractiveness.

The distilled portions were combined and extracted with
400 ml of hexane 3 times. The organic layers were combined,
washed with water, dried over anhydrous Na_2SO_4 and distilled.
The residual pale yellow oil was dissolved in a small quantity of hexane and charged on a column of alumina prepared
by pouring a hexane slurry of 50 g of material into a glass
tube, 4 cm dia. The column was developed with hexane followed by methanol. The hexane eluate (300 ml) and methanol

eluate (100 ml) on evaporation of the solvents yielded
440 mg of colorless oil and 150 mg of yellowish viscous
liquid, respectively. Both (A and B) residues were attractive to houseflies.

The residual solution after the above steam distillations
was still strongly attractive. Thus, the residual solution
(500 ml) was adjusted to pH 2 and repeatedly shaken with a
50 ml portion of chloroform. The chloroform layers were
separated, combined (ca. 500 ml), washed with water, dried
over Na_2SO_4 and evaporated under reduced pressure. The
viscous brown residue was dissolved in a small volume of
hexane and charged to a column prepared by pouring a hexane
slurry of 50 g of alumina into a glass tube, 4 cm dia. The
column was eluted in succession with 500 ml of hexane, 500
ml of benzene, 1000 ml of 1% methanol-benzene, 500 ml of 5%
methanol-benzene and finally 100 ml of methanol. The hexane
eluate on evaporation of the solvent yielded 75 mg of colorless viscous oil; 1% methanol-benzene eluate, 946 mg of pale
yellow viscous residue; and 5% methanol-benzene eluate,
113 mg of pale yellow viscous residue. These concentrates
(C, D and E, respectively) were also attractive to houseflies.

Since fraction D exhibited the strongest attraction, it
was examined further. The crude D concentrate typically
gave 3 spots on thin-layer chromatography: D_1, D_2 and D_3,
with R_F values of 0.14-0.16, 0.48-0.53 and 0.55-0.61 in 2%
methanol-benzene, respectively.

V. Properties of the Crystalline Compound D_3

Fraction D was further purified with silica gel chromatography. The crude D fraction (670 mg) was dissolved
in a small volume of benzene and charged to the column. The
column was developed with 1% methanol-benzene, and the eluate
was cut into fractions of 50 ml. The six 50 ml fractions
giving the D_3 spot alone on thin-layer chromatography were
combined, and evaporated to give a colorless waxy residue.
The waxy residues on repeated crystallization from ether-
methanol yielded 313 mg of colorless platelets, mp 22-23°C.

The compound D_3 was characterized as follows: an
apparent molecular weight, determined with Rast's method,
of 693, and $[\alpha]_D = \pm 0$ (1.79% in chloroform). IR and NMR
spectra suggested the presence of a long, straight chain,
an unsaturated fatty acid ester and a free secondary hydroxyl
group, as assigned in Fig. 1. These properties of D_3 were

compared with those of 1,3-diolein, which was prepared according to Daubert's method: acylation of 1-monotrityl glycerol with oleoyl chloride and acyl migration with dry HCl (Fig. 2). IR and NMR spectra of synthetic 1,3-diolein and of D_3 were identical. The identity of the fraction D_3 with 1,3-diolein was further supported by the products of hydrogenation and ozonolysis. D_3 absorbed 2.18 mols of hydrogen when reduced with 5% Pd-carbon in ethanol. The reduction product was identical with 1,3-distearin, prepared from trityl glycerol and stearoyl chloride on the basis of mixture melting point and infrared spectra. The ozonide obtained by passing ozone through an ice-cooled solution of D_3 in carbon tetrachloride was treated with aqueous ferrous sulfate and steam-distilled into a 2,4-dinitrophenylhydrazine-HCl solution. The hydrazone was identical with that from nonyl aldehyde (Volker and Soeren, 1961).

VI. Preparation of the Related Compounds

A. Preparation of the Compounds Related to 1,3-Diolein

1-Monoölein was prepared from glyceryl acetonide according to the method of Daubert et al. (1943), mp 30-33°C; 2-monoölein from benzylidene glycerol and oleoyl chloride followed by decomposition with boric acid in triethyl borate according to the method of Martin (1953), mp 28-30°C; 1,3-distearin from 1-trityl glycerol and stearoyl chloride, mp 76-77°C; 1,2-propanediol 1-monoöleate from the diol and oleoyl chloride, bp 196-198°C (2 mm Hg); glycerol 3-mono-chlorohydrin 1-monoöleate from glycerol monochlorohydrin and oleoyl chloride in quinoline, bp 273-275°C (1 mm Hg); and 3-methoxy glycerol 1-monoöleate by treating 3-mono-chlorohydrin 1-monoöleate with sodium methoxide in methanol, bp 202-206°C (1 mm Hg).

B. Preparation of Ethylene Glycol Monoöleate

Six ml of dry quinoline was gradually added to a solution containing 9 g of oleoyl chloride and 3 g ethylene glycol (about 0.8 molar excess) in 22 ml of dry chloroform. The reaction mixture was allowed to stand at room temperature for 1 hr, then refluxed for 4 hr. The reaction mixture was cooled, diluted with 50 ml of chloroform and successively washed with 2N H_2SO_4, cold water, 5% aqueous Na_2CO_3 and cold water. The solvent was removed, and the residue was dissolved in a small amount of benzene and placed on a column of 100 g alumina. After washing with benzene, the monoöleate was eluted with about 1000 ml of 5% v/v methanol-benzene. The solvent was evaporated and the residue distilled: bp 198-200°C (2mm Hg); yield 3.7 g (34%); anal. found C, 73.54%, H, 11.63%. Calculated for $C_{20}H_{38}O_3$: C, 73.57%; H, 11.73%.

C. Preparation of α,ω-Glycol Monoöleate

The following monoesters were prepared as above from the corresponding α,ω-glycols: propylene glycol monoöleate, bp 206-207°C (2 mm Hg), yield 23.5%; tetramethylene glycol monoöleate, bp 210-215°C (2 mm Hg), yield 16.3%; pentamethylene glycol monoöleate, bp 218-222°C (2 mm Hg), yield 28.5%; hexamethylene glycol monoöleate, bp 230°C (2 mm Hg), yield 15.7%; octamethylene glycol monoöleate, bp 284-285°C (3 mm Hg), yield 14.8%; decamethylene glycol monoöleate, mp 23°C, yield 25.7%.

VII. House Fly Attractiveness

A. Fraction D_3

Twenty mg of a crystalline preparation of D_3 was dissolved in 1 ml hexane and dropped into a petri dish. About 500 flies (age 10 days) were released in the testing room. Within 5 hr, 159 flies were in the dish containing the attractant, and only 2 (1 female and 1 male) were in the control dish.

One thousand flies (age 12 days) were released in the room and left without feeding for 10 hr. Then the petri dishes filled with powdered milk and a water supply (wick) was placed on the shelf. After nearly all the flies had gathered on the food, a petri dish containing 30 mg of D_3 was placed 60 cm from the food. The flies swarming over the food were gradually attracted to the test dish. After 24 hr, 662 flies (452 females and 210 males) were counted (Fig. 3).

B. 1,3-Diolein and Related Compounds

A number of 1,3-dioleins and related compounds were tested against a population of 1000 flies (age 5 days). Sample sizes were 1 mg, 10 mg, and 50 mg (Table I).

Table I

House Fly Attracting Activities of 1,3-Diolein and Related Compounds

Compounds	Chemical structure	BP (mm Hg)	MP °C	Numbers of flies attracted Expt.1	Expt.2
1,3-Diolein	$\begin{array}{l} \text{O} \\ \| \\ CH_2O\text{-}CC_{17}H_{33} \\ \mid \\ CHOH \\ \mid \\ CH_2O\text{-}CC_{17}H_{33} \\ \| \\ \text{O} \end{array}$		22-23	16^b	6^a
Triolein	$\begin{array}{l} \text{O} \\ \| \\ CH_2O\text{-}CC_{17}H_{33} \\ \mid \quad \text{O} \\ \qquad \| \\ CHO\text{-}CC_{17}H_{33} \\ \mid \\ CH_2O\text{-}CC_{17}H_{33} \\ \| \\ \text{O} \end{array}$	235-240 (15)		1^b	–

Table I--(Continued)

Compounds	Chemical structure	BP (mm Hg)	MP °C	Numbers of flies attracted Expt.1	Numbers of flies attracted Expt.2
1-Monoölein	$CH_2O-CC_{17}H_{33}$ (O) CHOH CH_2OH		30–33	33[a]	24[a]
2-Monoölein	CH_2OH $CHO-CC_{17}H_{33}$ (O) CH_2OH		28–30	19[a]	—
1,3-Distearin	$CH_2O-CC_{17}H_{35}$ CHOH $CH_2O-CC_{17}H_{35}$ (O)		76–77	0[b]	—

Table I--(Continued)

Compounds	Chemical structure	BP (mm Hg)	MP °C	Numbers of flies attracted	
				Expt.1	Expt.2
1,2-Propanediol 1-monooleate	CH3 / CHOH / CH2O-CC17H33 (O)	196-198 (2)		17a	—
Glycerol 3-monochlorohydrin 1-monooleate	CHCl / CHOH / CH2O-CC17H33 (O)	273-275 (1)		15b	—
3-Methoxyglycerol 1-monooleate	CH2OCH3 / CHOH / CH2O-CC17H33 (O)	202-206 (1)		16b	—

Table I--(Continued)

Compounds	Chemical structure	BP (mm Hg)	MP °C	Numbers of flies attracted Expt.1	Numbers of flies attracted Expt.2
Ethylene glycol monoöleate	$CH_2O-CC_{17}H_{33}$ (O) CH_2OH	198-200 (2)			22[a]
Ethylene glycol dioleate	$CH_2O-CC_{17}H_{33}$ (O) $CH_2O-CC_{17}H_{33}$ (O)			–	1[a]
Glycerol	CH_2OH $CHOH$ CH_2OH	125.5 (1)		1[c]	–
Methyl oleate	$CH_3O-CC_{17}H_{33}$ (O)			–	4[a]

a 1 mg per dish, after 16 hr.   b 10 mg per dish.   c 50 mg per dish.

199

C. α,ω-Glycol monoöleates

Samples of 1 mg each of a number of α,ω-glycol mono-öleates were tested. Results are presented in Table II.

Table II

House Fly Attracting Activities of α,ω-Glycol Monoöleate

C numbers in α,ω-glycols	Chemical structure	BP (mm Hg)	MP °C	Numbers of flies attracted after 16 hr (1 mg/dish)
2	$HO(CH_2)_2O-CC_{17}H_{33}$ (O)	198-200 (2)		31
3	$HO(CH_2)_3O-CC_{17}H_{33}$ (O)	206-207 (2)		33
4	$HO(CH_2)_4O-CC_{17}H_{33}$ (O)	210-215 (2)		20
5	$HO(CH_2)_5O-CC_{17}H_{33}$ (O)	218-222 (2)		28
6	$HO(CH_2)_6O-CC_{17}H_{33}$ (O)	230 (2)		29
8	$HO(CH_2)_8O-CC_{17}H_{33}$ (O)	284-285 (3)		5
10	$HO(CH_2)_{10}O-CC_{17}H_{33}$ (O)		23	2

D. Ethylene Glycol Monoöleate

Milk and a water supply (wick) were put at one end of
the shelf in the test room. A petri dish, 15 cm dia filled
with 300 ml of water was placed 1.5 m away, and 500 mg of
the ester in 300 ml of water was placed 1.5 m further.

Three days after 700 flies were released in the room,
311 were found in the test dish, 31 in the control dish, and
150 on the food.

Two cylindrical plastic traps, 22 cm high, 12 cm dia,
and provided with an inverted funnel-like bottom with a 3 cm
hole, were used. Petri dishes (9 cm dia) were attached to
the bottom at a distance of about 2 cm with cellophane tape.
About 10 ml of the ester were spread in the petri dish
attached to one trap, while the petri dish attached to the
other was left empty. About 100 ml of water containing 1 g
of detergent powder were poured into each of the traps to
retain the entering flies. The containers thus prepared
were hung 2 m apart on the wall of the test room in which
500 flies were released. Positions were rotated each day.
After 1 week, 106 flies were captured in the test traps and
26 in the control traps.

E. Saturated Monoacid Esters of Ethylene Glycol

Table III presents the results of tests involving 10
saturated monoacid esters of ethylene glycol.

Table III

House Fly Attracting Activities of Saturated Monoacid Esters of Ethylene Glycol

Fatty acids	Chemical structure	BP (mm Hg)	MP °C	Numbers of flies attracted after 16 hr 100·mga	10 mga	1 mga
Acetic	HOCH$_2$CH$_2$O-$\overset{O}{\overset{\|}{C}}CH_3$	184 (760)		2	14	12
Butyric	HOCH$_2$CH$_2$O-$\overset{O}{\overset{\|}{C}}C_3H_7$	111-113 (25)		1	5	27
Valeric	HOCH$_2$CH$_2$O-$\overset{O}{\overset{\|}{C}}C_4H_9$	76-78 (3)		5	12	32
Caproic	HOCH$_2$OH$_2$O-$\overset{O}{\overset{\|}{C}}C_5H_{11}$	94-95 (3)		6	7	37
Caprylic	HOCH$_2$CH$_2$O-$\overset{O}{\overset{\|}{C}}C_7H_{15}$	110-115 (3)		30	16	20

Table III—(Continued)

Fatty acids	Chemical structure	BP (mm Hg)	MP °C	Numbers of flies attracted after 16 hr		
				100 mga	10 mga	1 mga
Capric	HOCH₂CH₂O-CC₉H₁₉ (O‖)	150-151 (7)		62	46	31
Lauric	HOCH₂CH₂O-CC₁₁H₂₃ (O‖)	165-166 (5)		92	74	71
Myristic	HOCH₂CH₂O-CC₁₃H₂₇ (O‖)	180-182 (5)		1	21	7
Palmitic	HOCH₂CH₂O-CC₁₅H₃₁ (O‖)		48-49	10	22	42
Stearic	HOCH₂CH₂O-CC₁₇H₃₅ (O‖)		57	0	4	9

Table III—(Continued)

Fatty acids	Chemical structure	BP (mm Hg)	MP °C	Numbers of flies attracted after 16 hr		
				100 mg[a]	10 mg[a]	1 mg[a]
Oleic	$\overset{O}{\overset{\|}{HOCH_2CH_2O-CC_{17}H_{33}}}$	198–200 (2)		25	71	50
Control (water)	HOH			1	2	11

a Dose per dish.

VIII. References

Daubert, B. F., Fricke, H. H., and Longenecker, H. E. (1943).
    J. Am. Chem. Soc. 65, 2142.
Daubert, B. F., and Lutton, E. S. (1947). J. Am. Chem. Soc.
    69, 1449.
Martin, J. B. (1953). J. Am. Chem. Soc. 75, 5483.
Muto, T., and Sugawara, R. (1965). Agr. Biol. Chem. (Japan)
    29, 949.
Muto, T., Sugawara, R., and Mizoguchi, K. (1968). Agr. Biol.
    Chem. (Japan) 32, 624.
Takemoto, T., and Nakajima, M. (1964). Yakugaku Zasshi
    (Japan) 84, 1183, 1186, 1230, 1232.
Volker, F., and Soeren, O. (1961). Chem. Ber. 94, 1360.

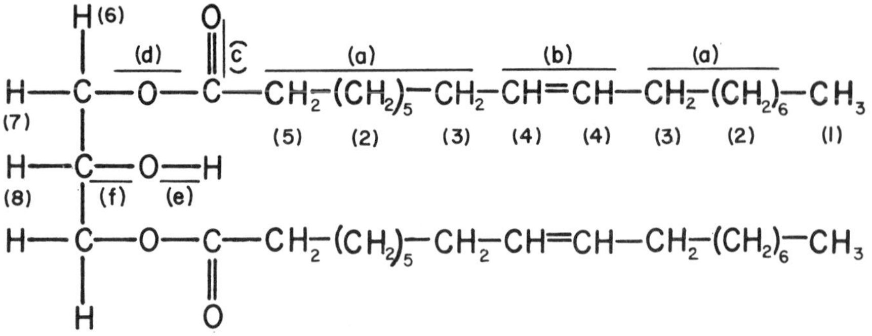

TOSHIO MUTO AND RYOZO SUGAWARA

NMR BANDS (ppm)	IR BANDS (cm^{-1})
(1) 0.90	(a) 721
(2) 1.28	(b) 1655
(3) 1.94	(c) 1743
(4) 5.28	(d) 1167
(5) 2.30	(e) 3480
(6) (7) 4.04 or 4.08	(f) 1091
(8) 4.04	

Figure 1.--Assignment of the NMR and IR bands characterizing fraction D$_3$.

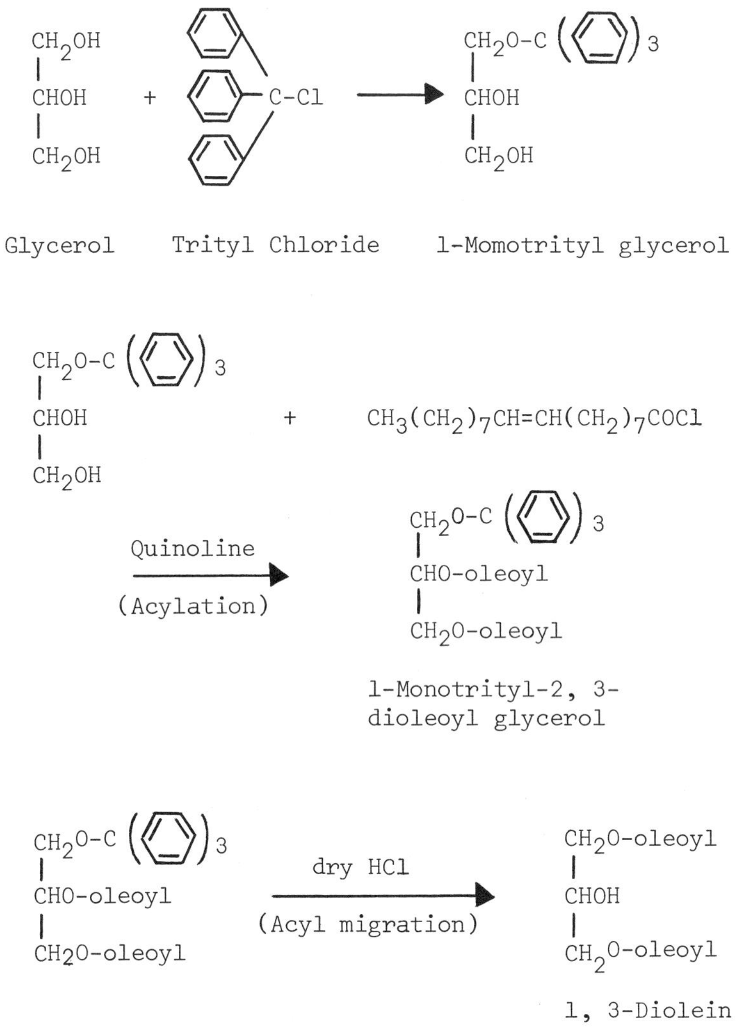

Glycerol   Trityl Chloride   1-Momotrityl glycerol

1-Monotrityl-2, 3-
dioleoyl glycerol

1, 3-Diolein

Figure 2.--Synthesis of 1,3-diolein.

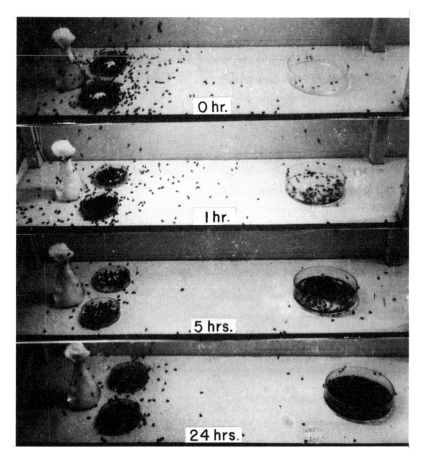

Figure 3.--House fly attractant activity of D₃.

STUDIES ON SEX PHEROMONES OF THE STORED GRAIN MOTHS

Minoru Nakajima

Department of Agricultural Chemistry
Kyoto University
Kyoto, Japan

Table of Contents

I. Introduction

There are various stored grain moths that cause serious damage to stored grain and its products. Fumigation has usually been applied for their control. Effective attractants or sex pheromones could provide a promising technique for preliminary surveys of moth populations and help establish a satisfactory control program.

The almond moth, Cadra cautella (Walker), is a widely distributed, serious economic pest of stored grain in Japan. The presence of a pheromone in the female almond moth has been reported by several investigators (Jacobson, 1965), but no attempt has been made to elucidate the chemical nature of the pheromone. We are studying the sex pheromone of the almond moth in cooperation with our colleagues at the Pesticide Research Institute and Dr. F. Takahashi, Entomological Laboratory, Kyoto University.

In this paper, we present a preliminary report of our experimental results.

II. Attempt to Isolate the Female Pheromone of Cadra cautella

A. Materials

Two strains of C. cautella were used in this experiment. Strain A was supplied by Japan Monopoly Corporation, Hatano Experimental Station, Kanagawa, and reared on rice bran at $25 \pm 1°C$ and 40 - 80% relative humidity. Strain B has been reared under a constant temperature (30°C) for more than 10 years in the Entomological Laboratory of the University. When eggs of this strain were collected from adult moths reared at 30°C and kept on rice bran for about 60 - 70 days at 20°C, only female moths emerged (Takahashi and Mutuura, 1964). The body weight and the number of emerging female moths were affected by the density of the eggs in rice bran (Fig. 1). From these data, the rearing conditions shown in Table I were adopted for collecting the female moths.

In the case of strain A, female moths were separated from male moths within 1 day after emergence. The collected female moths were dipped into methylene chloride and kept at -20°C until used.

CONTROL OF INSECT BEHAVIOR

Table I

Rearing Conditions of the Almond Moth

Strain	Eggs in rice bran	Temp.	Sex ratio	Numbers of females	Period from hatch to adult emergence
A	800/40g	25-28°C	1:1	280	33-40 days
B	1000/40g	20°C	♀♀ only	220	60-70 days

B. Procedure for the Bioassay

The assay for the pheromone was not affected by the
diurnal rhythm of male moths, and it could be carried out at
any time in the laboratory. Some 30 male moths were confined in a glass vial (11 cm dia x 7 cm deep) covered with
a glass plate having a hole (3 cm dia) in the center. A
glass rod was immersed 1 cm into a test solution and then
inserted through the hole into the glass vial. A positive
response was recorded if the male commenced wing movement
and hovered about the glass rod within 30 seconds. The
test solutions were prepared by successive tenfold dilution
of a 1 mg sample in 1 ml n-hexane, and were assayed beginning with the most dilute solution until a positive response
was observed.

In order to correlate the response of male moths with
days after their emergence, pupae were sexed, held in a vial
for adult emergence, and assayed every day after emergence.
Male moths are suitable for assay 2 - 3 days after emergence
(Table II).

The pheromone activities of virgin and mated female
moths were tested against a 1 female equivalent extract,
and it was shown that the pheromone activity decreased
rapidly to 1/100 or less within 5 hours after mating, whereas that of virgin female moths was unchanged.

C. Functional Group Tests for the Pheromone

Functional group tests on the crude extract suggest
without any doubt that the pheromone is an acetate of an
unsaturated alcohol (Table III).

211

Table II

Response of Male Almond Moths to the Pheromone
after Adult Emergence

Days after emergence	Concentration of the crude pheromone				
	1mg/ml	10^{-1}mg/ml	10^{-2}mg/ml	10^{-3}mg/ml	10^{-4}mg/ml
0 - 1	+	+	−	−	−
1 - 2	+	+	+	−	−
2 - 3	+	+	+	+	−
3 - 4	+	+	+	+	−
4 - 5	+	+	+	+	−

Table III

Functional Group Tests on the Crude Extract
of Almond Moth Pheromone

Reaction	Assay for the reaction product
Bromination	−
Debromination with zinc dust of the bromination product	+
Catalytic hydrogenation	−
Hydrolysis with dilute alkali	−
Acetylation of the hydrolysis product	+
Reduction with LiAlH$_4$	−

Table III--(Continued)

Reaction	Assay for the reaction product
Reduction with NaBH₄	+
Acetylation	+
Methylation with diazomethane	+
Reaction with 2,4-dinitro-phenylhydrazine	+

The disappearance of pheromone activity on bromination and also on catalytic hydrogenation indicated the presence of double bond(s) in the molecule, which was decisively confirmed by regeneration of activity on treatment of the bromination product with zinc dust. The activity disappeared on hydrolysis, but it was significantly regenerated on acetylation with acetic anhydride in pyridine. The activity also disappeared on reduction with lithium aluminum hydride. There was no change in the activity on the following reactions: acetylation with acetic anhydride in pyridine, methylation with diazomethane in ether, reduction with sodium borohydride, and treatment with 2,4-dinitrophenylhydrazine. These facts suggest that the pheromone is an acetate of an alcohol having one or more double bonds.

D. Isolation

The extract of 94,000 female moths was subjected to the procedures shown on the following page.

After separation of inactive fractions by precipitation from a chilled methanol and a chilled n-hexane solution, a yellow oil (7.3 g) was obtained. The gas-liquid chromatogram showed that the oil contained more than 7 components (Fig. 2). The gas chromatographic retention time of the pheromone was conveniently measured by the activity of male moths which were placed in a glass tube (1.5 cm x 15 cm)

Procedure for Isolation of the C. cautella Pheromone

Whole bodies of 94,000 female moths

    ├─ extracted with methylene
    │   chloride twice (6 1 + 2 1)

    ├─ filtered

Filtrate   Insoluble matters

    ├─ dried over anhydrous
    │   magnesium sulfate

    ├─ evaporated under
    │   reduced pressure

Yellowish brown oil (218 g)

    ├─ dissolved in methanol (400 ml)
    │   and kept at -20°C for 2 da

    ├─ decanted

Soluble fraction   Insoluble fraction

    ├─ evaporated under
    │   reduced pressure

Yellow oil (15 g)

    ├─ dissolved in n-hexane (18 ml)
    │   and kept at -20°C for 2 da

Soluble fraction   Insoluble fraction

    ├─ evaporated under
    │   reduced pressure

Yellow oil (7.3 g)

and attached to the end of the column, similar to the method
previously reported (Ignoffo et al., 1963; Shorey and Gaston,
1965a; Berger, 1966). The pheromone could be separated from
other inactive components by chromatography on silicic acid
(Mallinckrodt Chemical Works, 200 g, in a 3 cm x 64 cm
column) using n-hexane with an increasing ratio of benzene
as the eluting solvent (Fig. 3). Ten-gram fractions of
eluate were collected and the pheromone activity was found
in fractions nos. 511-630. It was a colorless oil weighing
about 40.6 mg, and was active at a concentration of 0.1 µg
/1 ml.

III. Species Specificity

It has been commonly assumed that a notable feature of
insect sex pheromones is their high specificity, and that in
most cases the female's scent attracts only males of the
same species. Recently, however, Shorey and Gaston (1965b)
found that insect sex pheromones were not always species
specific in the case of closely related noctuids. They
stated that, from a behavioral point of view, sex pheromones
of the cabbage looper, Trichoplusia ni (Huebner), and the
alfalfa looper, Autographa californica (Speyer), appeared to
be similar if not identical. Berger and Canerday (1968)
also found that the sex pheromone of cabbage looper was
active to Pseudoplusia includens (Walker), Rachiplusia ou
(Guenêe), and A. biloba (Stephens), which all belong to the
subfamily Plusiinae.

Barth (1937) reported that, in the case of Pyralidae,
males were attracted by female moths belonging to the same
subfamily, but not by those belonging to a different subfamily (Table IV).

While isolating the female sex pheromone of the almond
moth, we investigated species specificity, bearing in mind
Barth's findings.

A. Materials and Methods

The following 4 species of stored grain moths were used
in this experiment:
Phycitinae
  The almond moth, C. cautella.
  The Indian meal moth, Plodia interpunctella (Zeller).
  The Mediterranean flour moth, Anagasta kuehniella
  (Zeller).

215

Table IV

Responses of Male Moths to Female Moths in the Pyralidae[a]

Species of the female moths	Species of the tested male moths					
	Plodia interpunctella	Ephestia elutella	Anagasta kuehniella	Galleria mellonella	Achroea griella	Aphomia gularis
I						
Plodia interpunctella	+	−	−	−	−	−
Ephestia elutella	+	+	+	−	−	−
Anagasta kuehniella	+	+	+	−	−	−
II						
Galleria mellonella	+	−	−	+	+	−
Achroea griella	−	−	−	+	+	−
Aphomia gularis	−	−	−	−	−	+

I: Phycitinae; II: Galleriinae

a From Barth, 1937.

216

Pyralidinae
The meal moth, *Pyralis farinaris* (L.)

Pheromone extracts were prepared by soaking whole bodies of female moths in methylene chloride overnight and filtering to remove solid matter. The bioassay was the same as mentioned above.

The results shown in Table V support Barth (1937) except that male Mediterranean flour moths responded to the extract of female Indian meal moths. It was interesting that male Indian meal moths were highly responsive to extract of female almond moths. However, male almond moths were much less responsive to female Indian meal moth pheromone.

B. Functional Group Tests for the Pheromone

The functional groups and the gas chromatographic retention time of the pheromone of each species were tested. As shown in Tables VI and VII, there was no difference among the three species. These results indicate that the pheromones of the three species may be identical, or if not, may be closely related to each other in chemical structure.

Table V

Minimum Concentrations Required for Male Response
to Female Pheromones of Various Moth Species[a]

Pheromone extracted from	Species of male moth			
	C. cautella	P. interpunctella	A. kuehniella	P. farinaris
C. cautella	10^3ml	10^3ml	10ml	–
P. interpunctella	1ml	10^2ml	10ml	–
A. kuehniella	10^2ml	10ml	10^2ml	–
P. farinaris	–	–	–	10^2ml

[a] One female equivalent of pheromone presented in the indicated volumes of methylene chloride.

218

Table VI

Functional Group Tests of the Pheromones of Three Stored Grain Moths

Reaction	Species		
	C. cautella	P. interpunctella	A. kuehniella
Bromination	-	-	-
Hydrolysis	-	-	-
Acetylation	+	+	+
Reduction with LiAlH$_4$	-	-	-
Reduction with NaBH$_4$	+	+	+
Reaction with 2,4-dinitrophenylhydrazine	+	+	+

Table VII

Retention Times in Minutes of the Pheromones of Three Stored
Grain Moths Compared to Three Reference Compounds

Column conditions	C_{14}-OAc	C_{16}-OAc	C_{14}(unsat.)-OAc[a]	Pheromones[b]
1 m, 10% NGS, 200°C	7.9	14.9	–	9.6–12.4
1 m, 5% SE$_{30}$, 200°C	6.5	13.0	–	6.2– 9.0
2 m, 10% BDS, 170°C	30.0	61.6	35.6	46.0–52.0
2 m, 30% SG[c], 170°C	41.0	–	38.7	40.0–42.7

[a] 9-tetradecen-1-ol acetate.
[b] The pheromones of C. cautella, P. interpunctella, and A. kuehniella gave the same
   retention time.
[c] Silicone grease.

220

CONTROL OF INSECT BEHAVIOR

IV References

Barth, R. (1937). Zool. Jb. (Abtlg. Allg. Zool.
  Physiol.) 58, 297.
Berger, R. S. (1966). Ann. Entomol. Soc. Amer. 59, 767.
Berger, R. S., and Canerday, T. D. (1968). J. Econ. Entomol.
  61, 452.
Ignoffo, C. M., Berger, R. S., Graham, H. M., and Martin,
  D. F. (1963). Science 141, 902.
Jacobson, M. (1965). "Insect Sex Attractants" Wiley
  (Interscience), New York.
Shorey, H. H., and Gaston, L. K. (1965a). Ann. Entomol.
  Soc. Amer. 58, 604.
Shorey, H. H., and Gaston, L. K. (1965b). Ann. Entomol.
  Soc. Amer. 58, 600.
Takahashi, F., and Mutuura, A. (1964). Japanese J. Appl.
  Entomol. Zool. 8, 129.

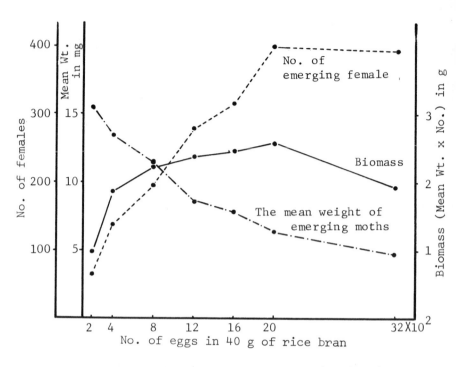

Figure 1.--Effects of almond moth egg density in rice bran.

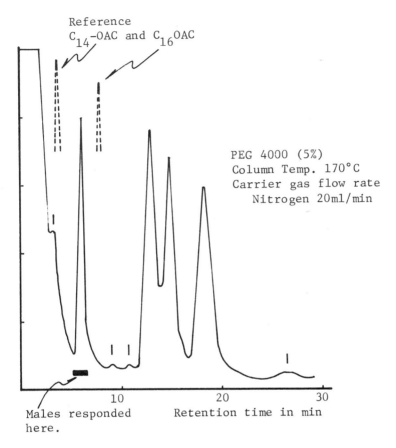

Figure 2.--Gas chromatogram of the crude almond moth pheromone.

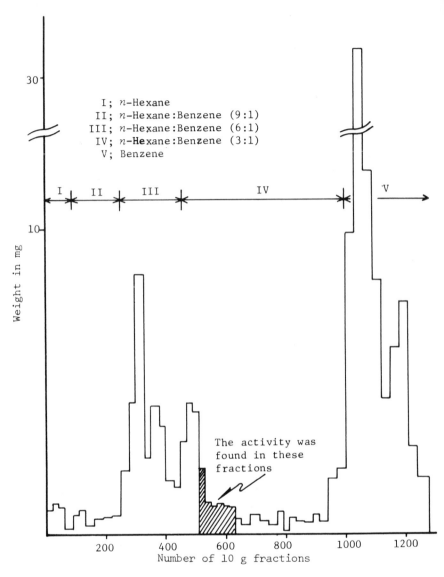

Figure 3.--Isolation of the almond moth pheromone by column chromatography.

INSECT ATTRACTANTS OF VEGETABLE ORIGIN,
WITH SPECIAL REFERENCE TO THE RICE STEM
BORER AND FRUIT-PIERCING MOTHS

Tetsuo Saito and Katsura Munakata

Faculty of Agriculture
Nagoya University
Nagoya, Japan

Table of Contents

I. Introduction

Insect attractants are known to be useful not only for
the survey of insect populations but also in their control.
Moreover, studies of naturally occurring attractants may
contribute to the elucidation of the host-selection mechanism and of varietal resistance in crops. Although considerable efforts have been made to isolate and characterize the
naturally occurring food lures responsible for insect attraction, only a few cases have been clarified.

In this paper, we will discuss our experiments on the
attractants of the rice stem borer and fruit-piercing moths
carried out at Nagoya University.

II. Attractant for the Rice Stem Borer

The rice stem borer, Chilo suppressalis Walker, is one
of the most destructive pests of the rice plant in Japan
and East Asian countries. In early summer, larvae hibernat-
ing in rice straw and stubble pupate and the moths emerge.
After mating, females migrate to paddy fields and lay their
eggs on rice plants. Newly hatched larvae immediately bore
into stems. The first generation larvae pupate in rice stems
and develop into moths in the late summer. The second generation larvae bore into rice plant stems and pass the winter
in the full grown larval stage inside infested stems or
stubble.

Some varietal differences in susceptibility to the
attack of the rice stem borer have been observed (Kawada,
1954; Seko and Kato, 1950; Wada, 1942). Serious infestation by the rice stem borer sometimes occurred in rice
plants cultured in paddy fields where a high level of nitrogen fertilizer was used or where 2,4-D was applied for control of weeds (Iyatomi and Sugino, 1951; Miyake, 1952). On
the other hand, rice plants which contained a higher level
of silica seemed more resistant (Sasamoto, 1953, 1955, 1957,
1961).

To account for the specificity shown by the stem borer
for the rice plant, an attractant may be present. In 1952
we started work on this problem.

A. Biological Assay

The longevity of the rice stem borer moth averages

7-12 days, depending upon its physiological state and environmental conditions. Moreover, rice stem borer moths are available only during 2 emergence periods. For these reasons, larvae were used as test animals during the isolation experiments.

The larvae were reared under aseptic conditions on a synthetic diet that contained no rice plant materials (Ishii and Urishibara, 1954). A test sample was dissolved in methanol, and 0.0012 ml of the solution was placed on small amounts of absorbent cotton at the bottom of 2 small test tubes (4 mm dia, 7 cm long). Two similar tubes treated with the same amount of methanol served as controls. The 4 tubes were taped in a cruciform pattern in a petri dish 18 cm dia, with their open ends pointing toward the center. Ten 30-day-old larvae were placed in the center of the petri dish after evaporation of the solvent, and held in the dark at 25°C for 12 hr. The larvae that crawled into the sample tubes were recorded as positive, those in the control tubes as negative, and those remaining outside the tubes as indifferent. This test was repeated 5 times, and if the differences in the mean number of insects found in the sample tubes were statistically highly significant, the sample was evaluated as positive.

B. Isolation of Active Principle

The air-dried immature rice plants were extracted with ethanol and the extract was re-extracted with ether. The active principle was found in the ether extract. This extract was separated into acidic, basic and neutral fractions. The neutral fraction was distilled under reduced pressure, and the distillate (< 140°C/2 mm Hg) was active. Since the activity was found in carbonyl compounds in preliminary tests, Girard-T reagent was utilized to separate carbonyl compounds from non-carbonyl compounds. Carbonyl compounds obtained were separated by chromatography. The scheme for isolation of the active principle is shown on the following page. The active fractions are in brackets. Each fraction was monitored by the biological assay described in the previous section.

The active principle was eventually purified by further chromatography, yielding 45 mg from 200 kg air-dried rice plants (the equivalent of about 2 tons of fresh plants). The active principle gave a single spot on a silica chromatostrip by detection with 2,4-dinitrophenylhydrazine. The

Isolation of the Active Principle of
the Rice Stem Borer
(Munakata and Okamoto, 1964)

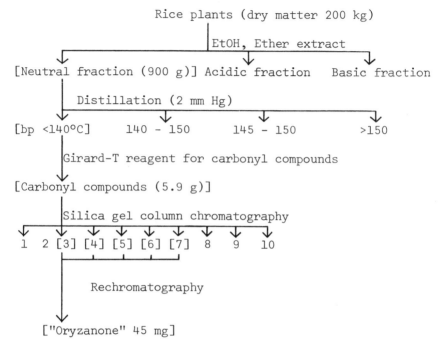

Rice plants (dry matter 200 kg)

|EtOH, Ether extract

[Neutral fraction (900 g)] Acidic fraction   Basic fraction

| Distillation (2 mm Hg)

[bp <140°C]   140 - 150   145 - 150   >150

|Girard-T reagent for carbonyl compounds

[Carbonyl compounds (5.9 g)]

|Silica gel column chromatography

1 2 [3] [4] [5] [6] [7] 8   9   10

Rechromatography

["Oryzanone" 45 mg]

effective minimal concentration of the attractant was
10^{-6} µg per test tube.

C. Identification of Attractant

The presence of a methyl ketone group in the molecule
of the attractant, $C_9H_{10}O$, was revealed by IR and a positive
iodoform reaction. Oxidation of the attractant by I_2 + KI
reagent liberated quantitatively iodofrom and ρ-methylben-
zoic acid. These results suggested that the attractant is
ρ-methyl acetophenone (I). Synthesized ρ-methyl acetophenone was identical with the natural substance by infrared

$$H_3C \text{---} \langle \bigcirc \rangle \text{---} COCH_3$$

(I)

and ultraviolet spectra, several chromatographic characteristics, and also by its biological activity. This attractive principle was named "Oryzanone", after the rice plant,
Oryza sativa L. (Munakata et al., 1959; Munakata and Okamoto,
1964).

D. Attractivity to Rice Stem Borer Moths

It is important to test for the attractivity of
ρ-methyl acetophenone for adult moths, because the host
selection takes place mostly in the adult stage. Two
methods, the orientation test and the oviposition test,
were employed. The attractivities were correlated with ρ-
methyl acetophenone contents in the extracts from rice
plants cultured under different conditions and from different varieties of rice plants.
Six varieties of rice plants were sown on soil in a
plastic pot, 25 x 40 x 12 cm, in a green house. One variety,
Aichi-asahi, was supplied with: (1) 0.01 g 2,4-D, (2) 15 g
silicic acid, (3) 3 g ammonium sulfate. After 1 month,
20 g samples of rice plants were extracted with methanol
using a Soxhlet extractor. The extracts were diluted with
methanol to 250 ml for biological tests and chemical analyses.
The orientation test was conducted using a covered
plastic box, 23 cm long, 30 cm wide, and 6 cm deep, with 2
holes--one leading to a tube carrying the odor, and the
other to a control air stream. The temperature was 28°C.
Circulation was maintained by withdrawing air (5 l/min)
through 2 additional holes as shown in Fig. 1. Fifteen
moths were released into the box, and the number exhibiting
a positive response was counted at regular intervals using
minimal light (< 1 lux).
The oviposition test was conducted as follows: a filter paper, Toyo No. 2, 15 cm dia, was impregnated with a 1
ml methanol solution of the extracts. After evaporation of
the solvent, 2 treated filter papers were taped in a plastic
box in which 50 female moths had been released. Preference
tests between the 2 filter papers for oviposition were
carried out under the same conditions as described for the
orientation tests. These experiments were started in the
evening, and after 18 hr, egg masses on the filter papers
were counted.
The content of ρ-methyl acetophenone was determined by

gas chromatography. More attractivity was found in rice
plants with a higher ρ-methyl acetophenone content, i.e.,
those plants to which high levels of nitrogen or 2,4-D had
been applied (Table I). On the other hand, rice plants

Table I

Relationship Between Attractivity to the Rice Stem Borer
Moth and ρ-Methyl Acetophenone Content in Extracts of Rice
Plants Treated with High Nitrogen Fertilizer,
Silicate, and 2,4-D

Treatment	Orientation test (moths)	Oviposition test (egg masses)	Content of ρ-methyl acetophenone in fresh plants (µg/g)
High nitrogen	251	11	0.72
Control	199	4	0.16
Silicate	202	2	0.09
Control	248	13	0.16
2,4-D	249	19	1.36
Control	201	6	0.16

supplied with silicate showed a lower content of ρ-methyl
acetophenone, resulting in a decreased attractivity. There
was no consistent relationship between moth attractivity
and content of ρ-methyl acetophenone in different varieties
of rice plants (Table II).

Thus, we suspect that the preference for rice plant
varieties may involve not only ρ-methyl acetophenone but
also other attractants (Kono and Saito, 1963).

III. Attractant for Fruit-Piercing Moths

Fruit-piercing moths, such as the akebia leaf-like
moth, Adris tyrannus amurensis Staudinger, the reddish
oraesia, Oraesia excavata Butler, the small oraesia, O.
emarginata Fabricius, and others, are important insect pests

Table II

Relationship Between Attractivity to the Rice Stem Borer
Moth and ρ-Methyl Acetophenone Content in Extracts of
Several Varieties of Rice Plants

Variety	Orientation test (moths)	Oviposition test (egg masses)	Content of ρ-methyl acetophenone in fresh plants (µg/g)
Kameno	261	28	0.88
Senbon-asahi	139	11	0.91
Aichi-asahi	240	11	0.16
Norinmochi No.1	210	12	1.02
Kurotaikyu	254	19	2.20
Norin No.8	196	7	1.36

attacking many kinds of orchard fruits. They are distributed not only in Japan but also in many tropical countries (Golding, 1945; Nakajima and Shimizu, 1950).

In the twilight, they fly long distances from mountain shrubs to orchards and suck juices by piercing the fruit with a chitinized proboscis. In the early morning they fly back to resting places. Fruits attacked sometimes were infected with microorganisms. The larvae are widely distributed in bushes, and feed on vine plants such as Cocculus trilobus Dc.

Early harvest of fruits, bagging fruits with polyethylene bags, and illumination of orchards are recommended for reducing damage (Matsuzawa, 1961; Nomura, 1967).

A large number of moths was attracted to traps, containing ripe fruit, which had been set on the edge of orchards. This finding suggested that ripe fruit contains an attractant. The following studies were undertaken to isolate the attractant from fruit and to utilize it as a control method.

A. Biological Assay

The following methods were evaluated: proboscis response test, piercing test, large Y-tube type olfactometer
test and trap cage test. The trap cage test was considered
to be the most suitable method. Plastic traps, 15 cm dia
and 25 cm long, containing the various fractions in 2 g of
liquid paraffin, were hung inside a screened cage, 3.6 x
3.6 x 2.7 m, in which a number of fruit-piercing moths had
been released. The number of moths caught in the traps containing candidate samples gave an estimate of attractivity
of the samples. The attractivity of white peach juice is
shown in Table III (Saito, 1960).

B. Isolation and Attractivity

The first attempt at collection and isolation was conducted on an ether extract of peach juice. However, the
extract did not show strong attractivity to the moths (Saito,
1961). The next attempt at isolating the attractant employed a modification of the method described by Yamamoto
(1963). About 120 kg of ripe grapes were placed in a 180 l
drum, and a stream of air, dried by passing through conc.
H_2SO_4 and NaOH, was passed through the drum (10 ml/min).
The air was then introduced into a 5 l flask chilled in an
electric freezer at -20°C, and then into a 2 l flask immersed
in an alcohol-dry ice bath at -70°C.
An aqueous condensate (45 l, pH 3.5) was obtained from
1080 kg of "Niagara" grapes over a 24 day period. An aliquot
was saturated with sodium chloride and extracted with ether.
The extract was concentrated to 4.5 l at 40-50°C to give a
stock solution.
A 0.5 l portion of the stock solution was separated
into acidic, basic and neutral fractions. The neutral fraction, which showed strong attractivity to the moths, was
treated with Girard reagent to separate carbonyl compounds
by the method of Gaddis et al. (1964). The noncarbonyl
faction showed strong attractivity.
The stock solution of 0.5 l was also fractionated by
chromatography on silicic acid (Mallinckrodt, 100 mesh),
eluted with pentane, 10, 30 and 50% ether in pentane, ether,
and methanol, successively. The strongest activity was
found in the 50% ether in pentane fraction. Gas chromatography (Fig. 2) of this fraction showed that the active

Table III

Attractivity of White Peach Juice to Fruit-Piercing Moths

Juice (ml/trap)	3 traps/cage		2 traps/cage		1 trap/cage	
	Number of moths trapped	Per cent	Number of moths trapped	Per cent	Number of moths trapped	Per cent
20	20	33	16	20	6	10
2	1	2	7	9	1	2
0	0	0	0	0	0	0

233

principle was eluted from 0 to 14 min. After saponification
of this fraction, peak 1 disappeared and the saponified
material was not attractive. This suggests that the active
component was attributable to peak 1 (Munakata et al., 1967).

Acknowledgments

The authors thank Professor Kisabu Iyatomi, Nagoya
University, for his valuable advice. The contributions of
the following co-workers are gratefully acknowledged:
Messrs. Hachiro Honda, Akio Miyazaki and Nobuo Tsuihiji.

CONTROL OF INSECT BEHAVIOR

IV. References

Gaddis, A. M., Ellis, R., and Currie, G. T. (1964). Food Res. 29, 6.
Golding, F. D. (1945). Bull. Entomol. Res. 36, 181.
Ishii, A., and Urishibara, H. (1954). Bull. Nat. Inst. Agr. Sci. 4, 109.
Iyatomi, K., and Sugino, T. (1951). Plant Protect., Japan 5, 120.
Kawada, A. (1954). Rept. 5th Meet. Internat. Rice Comm. Working Party Rice Breed., 151.
Kono, T., and Saito, T. (1963). Paper presented at the Annual Meeting of the Japanese Society of Applied Entomology and Zoology, Tokyo.
Matsuzawa, H. (1961). Special Report of the Lab. Appl. Entomol. Fac. Agr. Kagawa Univ. 1.
Miyake, T. (1952). Plant Protect., Japan 6, 311.
Munakata, K., Saito, T., Ogawa, S., and Ishii, S. (1959). Bull. Agr. Chem. Soc. Japan 23, 64.
Munakata, K., and Okamoto, D. (1964). In "The Major Insect Pests of the Rice Plant". Johns Hopkins Press, Baltimore, Maryland.
Munakata, K., Saito, T., and Miyazaki, A. (1967). Paper presented at the Annual Meeting of the Japanese Society of Applied Entomology and Zoology, Tokyo.
Nakajima, S., and Shimizu, K. (1950). Ōyō-Kontyū 6, 30.
Nomura, K. (1967). Jap. J. Appl. Entomol. Zool. 11, 21.
Saito, T. (1960). Abstract for 4th Symposium Jap. Soc. Appl. Entomol. Zool., 30.
Saito, T. (1961). Abstract for 5th Symposium Jap. Soc. Appl. Entomol. Zool., 1.
Sasamoto, K. (1953). Ōyō-Kontyū 9, 108.
Sasamoto, K. (1955). Ōyō-Kontyū 11, 66.
Sasamoto, K. (1957). Botyu-Kagaku 22, 159.
Sasamoto, K. (1961). Proc. Fac. Liberal Arts and Educ. 3, 3.
Seko, H., and Kato, I. (1950). Proc. Crop. Sci. Soc. Japan 19, 201.
Wada, E. (1942). Kagaku 12, 441.
Yamamoto, R. (1963). J. Econ. Entomol. 56, 119.

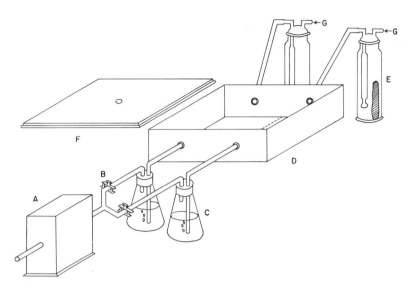

Figure 1.--The olfactometer for orientation test of the rice
stem borer moth. A: suction pump; B: screw clamp; C:
bubbling bottle for air stream control; D: reaction
box; E: odor sample; F: cover; G: fresh air.

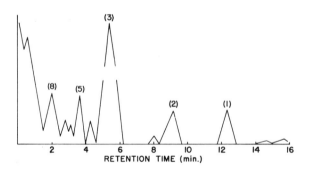

Figure 2.--Gas chromatogram of 50% ether in pentane fraction
from silicic acid chromatography. Column = PEG-6000,
3 mm x 225 cm; N_2 gas = 60 ml/min; He gas = 50 ml/min;
50-170°C, 6°C/min.

THE CHEMISTRY OF ATTRACTANTS FOR CHRYSOPIDAE
FROM Actinidia polygama Miq.

Takeo Sakan, Sachihiko Isoe and Suong Be Hyeon

Department of Chemistry
Osaka City University
Osaka, Japan

Table of Contents

Table of Contents

CONTROL OF INSECT BEHAVIOR

I. Introduction

"Matatabi" <u>Actinidia polygama</u> Miq., Actinidiaceae, a vine-like plant widely distributed in mountainous areas of Japan, has long been of interest because of its peculiar excitatory effect on cats and other Felidae, and its utilization in folk medicine and as an analeptic. From this plant, Sakan <u>et al.</u> (1959, 1960, 1965, 1967) have isolated, identified and synthesized several cyclopentanoid monoterpenes to which Felidae respond. These are the matatabilactones (iridomyrmecin, isoiridomyrmecin, dihydronepetalactone, isodihydronepetalactone and neonepetalactone), actinidine, C-11 lactones (actinidiolide and dihydroactinidiolide) and phenylethyl alcohol.

Ishimori (1931), Yano (1931) and Takagi (1932) have reported the interesting observations that male adults of lacewings, <u>Chrysopa septempunctata</u> Wesmael (Chrysopidae), are highly attracted by A. <u>polygama</u>; that is, they swarm on the leaves and fruits. Ishii (1964) later showed that the neutral extract of the fruits was attractive to the insect. We have continued the extensive investigation on the components of A. <u>polygama</u> and have isolated several active attractants for lacewings.

II. Isolation of Attractants

Preliminary tests showed that the active components were present in a neutral volatile fraction of a methanolic extract of leaves. When a thin-layer chromatogram of this fraction was exposed overnight to adult male insects, the active zone was delineated by biting marks.

A. Isolation from Galls

The crude neutral fraction from volatile oil obtained by the steam distillation of a methanol extract of galls was distilled at reduced pressure. The main fraction (80-130°/2 mm) was treated with hot alkaline solution to remove lactones, and was fractionated by column chromatography on silica gel, followed by preparative gas chromatography. The following active compounds were identified: iridodiol, matatabiol, 5-hydroxymatatabiether, 7-hydroxydihydromatatabiether, and allomatatabiol. However, these compounds were less active than the active zone of the preliminary thin-layer chromatogram.

B. Isolation from Leaves

The steam distillate of a methanol extract of leaves
was distilled at reduced pressure, and the fraction that
distilled at 35-170°/0.18 mm was treated with alkali to
remove lactones and chromatographed on alumina. Each active
fraction was further purified by gas-liquid and thin-layer
chromatography. Beside matatabiol and 5-hydroxymatatabi-
ether a small amount of an extremely active alcohol, named
neomatatabiol, was isolated. Its Rf value was completely
consistent with that of the active zone of the thin-layer
chromatogram.

C. Large Scale Isolation of Neomatatabiol from Leaves

Leaves were steam distilled and the distillate was
treated with alkali. By repeated column chromatography of
the neutral fraction on alumina, several grams of crude
active alcohols were obtained from 1000 kg of fresh leaves.
In this case a small amount of dehydroiridodiol, which seems
to be a biogenetically important intermediate, was also
isolated. By repeated column chromatography of the crude
active alcohols, small amounts of pure neomatatabiol and
very small amounts of isoneomatatabiol were isolated.

III. Biological Test

A filter paper impregnated in a circle of ca. 1 cm
dia with a very dilute solution of a known concentration of
weighed sample was exposed to natural and caged populations
of insects after sunset. After several minutes, lacewings
began to swarm on the test paper. It was considered to be
attractive if biting marks on the paper could be detected
by the naked eye.

The monoterpene alcohols isolated from A. polygama were
found to be attractive only for the male adults of C.
septempunctata and C. japana Okamoto down to the levels of
1 μg of iridodiol, 5-hydroxymatatabiether, 7-hydroxydihydro-
matatabiether and allomatatabiol, 10^{-3} μg of matatabiol and
dehydroiridodiol, and 10^{-6} μg of neo- and isoneomatatabiol.

IV. Elucidation of Attractant Structures

A. Iridodiol and Dehydroiridodiol

As previously reported by Sakan et al. (1964), iridodiol was found to be a mixture of structures (Ia), (Ib) and (Ic) by gas chromatographic comparison with authentic samples. Structure (II) was assigned for dehydroiridodiol as a result of identity of the infrared spectrum and the gas-chromatogram of dehydroiridodiol with those of an authentic sample obtained by the degradation of matatabiether.

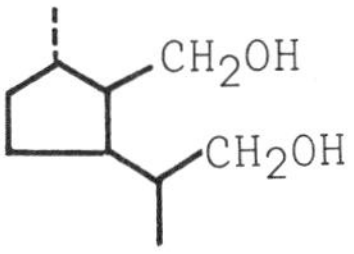

(Ia) 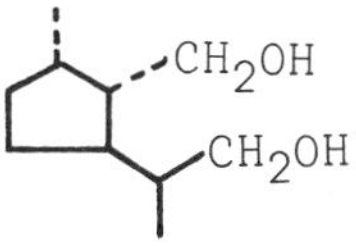   (Ib) 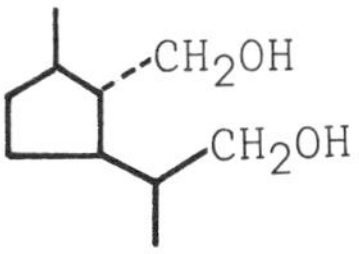   (Ic) 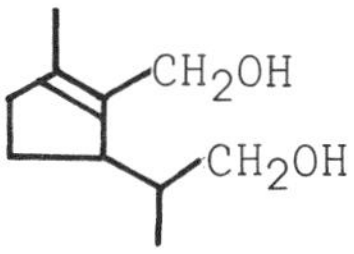   (II)

B. 5-Hydroxymatatabiether

5-Hydroxymatatabiether (III) was obtained from both leaves and galls, especially the latter, as the major component of the alcohol fraction. 5-Hydroxymatatabiether

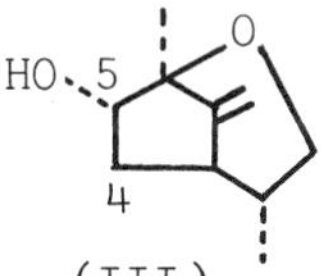

(III)

(colorless liquid, $C_{10}H_{16}O_2$, bp 100-103°/1 mm, $[\alpha]_D$ -134.5°) decolorizes bromine solution and exhibits infrared (IR) absorption at 3400 and 1045 cm^{-1} (OH), 3080, 1680 and 895 cm^{-1} (=CH$_2$), and 1130, 1110, 1090 and 1020 cm^{-1} (ether). The NMR spectrum of 5-hydroxymatatabiether shows a doublet signal of methyl at 9.1τ (3H, J=6 cps), a singlet signal of methyl attached to tertiary carbon atom adjacent to oxygen at 8.7τ, a singlet signal of OH at 6.5τ, a multiplet signal of CH-CH$_2$-O centered at 6.5τ constituting the AB part of an ABX pattern (J_{AX}=11.8 cps, J_{BX}=6.2 cps and J_{AB}=11.3 cps), two doublet signals of methine (CHOH) at 5.77τ and a singlet signal of terminal methylene at 5.1τ. Catalytic hydrogenation of 5-hydroxymatatabiether using platinum oxide in ether afforded the dihydro-derivative (IV). The NMR spectrum of

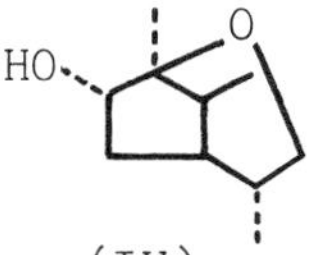

(IV)

241

dihydroether (IV) is similar to that of 5-hydroxymatatabi-
ether, except that the signal at 5.1τ (=CH$_2$) was replaced by
a doublet methyl at 9.2τ (3H, J=6 cps). The similarities of
IR and NMR spectra of 5-hydroxymatatabiether to those of
matatabiether (V), whose structure has already been established, strongly suggested that 5-hydroxymatatabiether
would have the same skeleton as matatabiether. This was

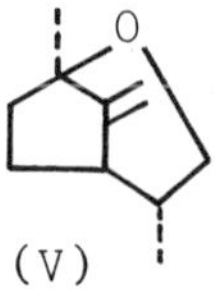

(V)

further demonstrated by lithium aluminum hydride reduction
of the tosylate of the dihydroether (IV) to dihydromatatabiether (VI). In the mass spectra of (IV) and (VI), the

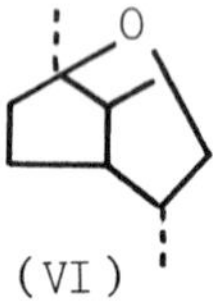

(VI)

fragmentation patterns below mass 123 were similar. Catalytic hydrogenation of (III) over Pd-C gave the dihydroether
(IV) and 9-hydroxynepetane (VII). Since the stereochemistry
of matatabiether has already been established, 5-hydroxy-
matatabiether is represented by the formula (III).

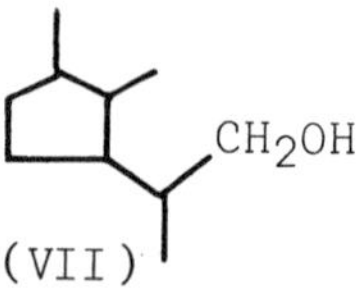

CH$_2$OH

(VII)

C. 7-Hydroxydihydromatatabiether

The NMR spectrum of 7-hydroxydihydromatatabiether
(colorless liquid, C$_{10}$H$_{18}$O$_2$, [α]$_D$-16°) bears a close resemblance to that of dihydromatatabiether (VI) except that a
doublet methyl signal at 9.0τ (3H, J=6 cps) of dihydromatatabiether is replaced by a doublet signal of CH-CH$_2$-O at 6.18τ
(J=6.8 cps) and a singlet signal of the hydroxyl proton at
7.25τ. From the above spectroscopic evidence, structure
(VIII) was assigned for 7-hydroxydihydromatatabiether. This
assignment was supported by the comparison of the mass spectrum of (VIII) with that of dihydromatatabiether (VI).

242

(VIII)

D. Allomatatabiol

Allomatatabiol (IX) (colorless liquid, $C_{10}H_{16}O_2$, $[\alpha]_D$-91°) decolorizes bromine solution. The IR and NMR spectra of allomatatabiol bear a close resemblance to those of allomatatabiether (X) previously obtained by the treatment of matatabiether (V) with acids, except for hydroxyl absorption at 3400 and 1050 cm^{-1}, and a signal for the hydroxyl proton at 7.05τ and the CHOH proton adjacent at 5.87τ. Thus the structure of allomatatabiol was deduced to be a hydroxyl allomatatabiether in which the hydroxyl group is located at C-4 or C-5; the latter position is more favorable in terms of biogenesis. This was demonstrated by the conversion of 5-hydroxymatatabiether (III) to allomatatabiol (IX) with 4N hydrochloric acid.

(III)   (IX)   (X)

E. Matatabiol

Matatabiol (XI) (colorless liquid, $C_{10}H_{16}O_2$, $[\alpha]_D$-16.2°, p-nitrobenzoate, mp 78-80°) decolorizes bromine solution. The IR spectrum of matatabiol exhibits absorption at 3040 cm^{-1} (ν_{C-H}), 1650 cm^{-1} ($\nu_{C=C}$), 865 and 820 cm^{-1} (δ_{C-H}), 3420 cm^{-1} (ν_{OH}) and 1040 cm^{-1} (ν_{C-O}). The NMR spectrum of matatabiol shows a doublet signal of methyl at 9.12τ (3H, J=6 cps), a broad singlet signal of a methyl on a double bond at 8.40τ (3H), a signal of hydroxyl proton at 7.95τ, a triplet signal of methylene at 6.37τ (2H, triplet of AB type J=11.7 cps), a multiplet signal of methylene adjacent to an oxygen atom between 6.8τ and 6.2τ constituting AB part of ABX pattern (J_{AX}=11.2 cps, J_{BX}=5.3 cps and J_{AB}=7.5 cps), and a signal of an olefinic proton at 4.5τ (1H, broad singlet).

Matatabiol absorbs one mole of hydrogen to give dihydromatatabiol. The NMR spectrum of dihydromatatabiol no

243

longer contains signals of the olefinic proton and the methyl group on the double bond; these are replaced by a doublet signal of methyl at 9.07τ, overlapped by another doublet signal of methyl.

It follows therefore that dihydromatatabiol has structure (XII), and matatabiol has structure (XI). The exhaustive catalytic hydrogenation of matatabiol over Pd-C yielded 9-hydroxynepetane (VII). This also supports the structure (XI) for matatabiol.

(XI)   (XII)

F. Neomatatabiol and Isoneomatatabiol

Neomatatabiol (XIII) ($C_{10}H_{18}O_2$, colorless liquid, bp 95°/5 mm, $[\alpha]_D$ +21.3°) gives a positive Tollens test, which indicates the presence of a potential aldehyde group. The IR spectrum of neomatatabiol showed a hydroxyl band. Since neomatatabiol affords dihydronepetalactone on chromic acid oxidation, the structure (XIII) was assigned for neomatatabiol. This assignment was further supported by the NMR spectrum; two doublet methyl signals at 9.20τ and 9.02τ (6H, J=6 cps, respectively), multiplet signals of CH-CH$_2$-O-between 6.3τ and 7.0τ constituting AB part of an ABX pattern (J_{AX}=10.9 cps, J_{BX}=4.1 cps and J_{AB}=11.2 cps), and a doublet signal of methine between oxygen atoms at 5.7τ (1H, J=8 cps).

The structure (XIII) was eventually proved by the mild reduction of dihydronepetalactone (XIV) to neomatatabiol with lithium aluminum hydride. The reduction of isodihydronepetalactone (XVI) with lithium aluminum hydride gave mainly isoneomatatabiol (XV), the IR spectrum of which differs from that of neomatatabiol in the fingerprint region, but the mass spectrum of isoneomatatabiol was almost identical with that of neomatatabiol. The NMR spectrum of isoneomatatabiol (XV) shows the presence of two doublet methyls at 9.20τ and 8.95τ, multiplet methylene of a -O-CH2CH- group centered at 6.5τ, and a doublet at 4.92τ (J=2 cps) attributed to a -O-CH-O group. From the above chemical and spectroscopic evidence, the structure of isoneomatatabiol was represented by (XV).

(XIII)   (XIV)   (XV)   (XVI)

V. Structural Relationship Between
Attractant and Defensive Agent

The ant lactones, iridomyrmecin (XVII) and isoiridomyrmecin (XVIII) isolated from Iridomyrmex humilis (Mayr)
(Pavan, 1950; Fusco et al., 1955), I. nitidis Mayr (Cavill
et al., 1956) and Dolichoderus scabridus (Roger) (Cavill
and Hinterberger, 1960), have attracted considerable attention because of their use by the ants as defensive agents
against preying insects and as possible means of communication. These lactones have also been found in A. polygama
accompanied by dihydronepetalactone (XIV), isodihydronepetalactone (XVI) and neonepetalactone (XIX). Iridodial
(XX) and dolichodial (XXI), other defensive agents isolated
from several species of Iridomyrmex ants (Cavill et al.,
1956; Cavill and Hinterberger, 1960; Trave and Pavan, 1956)
and Anismorpha buprestoides (Stoll) (Phasmidae) (Meinwald
et al., 1962), also have the same structure as the afore-
mentioned attractants.

It is most remarkable that the chemical structures of
potent attractants for lacewings correspond to the reduced
forms of these defensive lactones and dialdehydes.

(XVII)   (XVIII)   (XIX)

(XX)   (XXI)

245

VI. Defensive Agent of Lacewings

It is well known that both sexes of lacewings, when disturbed, eject a skatole-like offensive odor. This secretion of lacewings may be the defensive agent which is effective against predatory insects and birds. To identify the defensive agent, we prepared an ethanol extract of 10,000 adult lacewings, which was then steam distilled. From the distillate, skatole was isolated and identified by gas chromatography using a HU2000 Golay column.

VII. Agricultural Significance

Both larval and adult stages of lacewings are predatory and some species are valuable aids to man in the natural control of insects such as plant-lice, scales and thrips, which are pests in farm, garden and orchard. It may be possible to use synthetic lacewing attractants for the local control of populations of certain insects as well as for survey studies of these predators.

VIII. References

Cavill, G. W. K., Ford, D. L., and Locksley, H. D. (1956).
    Austral. J. Chem. 9, 288.
Cavill, G. W. K., and Hinterberger, H. (1960). Austral. J.
    Chem. 13, 514.
Fusco, R., Trave, R., and Vercellone, A. (1955). Chim. e
    Ind. 37, 251.
Ishii, S. (1964). Jap. J. Appl. Entomol. Zool. 8, 334.
Ishimori, N. (1931). Kontyu 5, 96.
Meinwald, J., Chadha, M. S., Hurst, J. J., and Eisner, T.
    (1962). Tetrahedron Letters No. 1, 29.
Pavan, M. (1950). Ricerca Sci. 20, 1853.
Sakan, T., Fujino, A., Murai, F., Butsugan, Y., and Suzui, A.
    (1959). Bull. Chem. Soc. Japan 32, 1156.
Sakan, T., Fujino, A., and Murai, F. (1960). Nippon Kagaku
    Zasshi 81, 1320, 1324, 1327.
Sakan, T., Isoe, S., Hyeon, S. B., Ono, T., and Takagi, I.
    (1964). Bull. Chem. Soc. Japan 37, 1888.
Sakan, T., Isoe, S., Hyeon, S. B., Katsumura, R., Maeda, T.,
    Wolinsky, J., Dickerson, D., Slabaugh, M., and Nelson,
    D. (1965). Tetrahedron Letters No. 46, 4097.
Sakan, T., Isoe, S., and Hyeon, S. B. (1967). Tetrahedron
    Letters No. 17, 1623.
Takagi, G. (1932). Zool. Mag. 44, 477.
Trave, R., and Pavan, M. (1956). Chim. e Ind. 38, 1015.
Yano, M. (1931). Kontyu 5, 96.

SEX PHEROMONES OF LEPIDOPTERA

H. H. Shorey

Department of Entomology
University of California
Riverside, California

Table of Contents

H. H. SHOREY

Table of Contents

CONTROL OF INSECT BEHAVIOR

I. Introduction

This review is concerned with those Lepidoptera (generally the moths) that use female sex pheromones for communication over some distance. The butterflies, superfamily Papilionoidea, have mainly evolved a visual system for distance communication between the sexes and are not considered here.

Lepidopterous sex pheromones have received a great deal of attention during recent years. This is due in part to economic reasons; the larvae of certain species of Lepidoptera represent some of our most serious agricultural pests. Difficulties that arise in controlling these pests with conventional techniques, mainly insecticidal, have caused an increased demand for more research into alternate means of control, including behavioral manipulation by sex pheromones (Shorey and Gaston, 1967). Also, with the development of modern techniques and instrumentation, sex pheromones of Lepidoptera have offered an exciting substrate to chemists interested in working with submilligram quantities of chemicals and to electrophysiologists interested in exploring mechanisms of olfaction.

A number of reviews have appeared in recent years summarizing and analyzing the literature in this field (Jacobson, 1965, 1966; Karlson and Butenandt, 1959; Shorey and Gaston, 1967; Shorey et al., 1968; Wilson, 1963). Therefore, instead of a comprehensive review, this article is an attempt to integrate recent developments in certain important areas and to speculate about future research.

II. Quantitative Bioassays

As the name implies, a bioassay is an assay for the presence or the quantity of a substance, with the detector being a living organism or some portion of that organism (such as antennal sensory cells). A bioassay is the basic tool used in assessing biological activity during chemical and behavioral studies of pheromones.

A. Development of a Bioassay

Let us consider the events that probably occur in nature, as the pheromone molecules are released from the gland of a female moth, and impinge on the antennal receptor

251

surfaces of a sexually responsive male of the same species.
The behavioral steps that ensue will be considered, using
the cabbage looper moth, Trichoplusia ni (Hübner)
(Noctuidae) as an example (Shorey, 1964). Consider that the
time is between midnight and dawn, the normal time for
mating in this species. The pheromone is carried by air
currents from the female to the male, which is in the
normal "resting" position, with its wings folded back roof-
like over the body and its antennae held back along or
under the leading margins of the wings. The first noticeable sign that the male is stimulated occurs when the male
brings the antennae forward. The wings are then extended
and vibrated, with the vibrations increasing in amplitude;
presumably the vibrations serve to warm the thorax temperature to the critical level necessary for flight (Heath and
Adams, 1967). After the warm-up period, which may last only
a few seconds in a warm environment or after the moth has
recently settled, or may persist for several minutes under
cooler conditions, the moth takes flight and orients itself
in an upwind direction. The physiological means by which
the male maintains its upwind direction is subject to a
number of conflicting theories (See Section III). It is
possible that many males never reach the female. Assuming
that our specimen does, a further sequence of behavior
ensues. The female is usually clinging to a vertical surface and vibrating her wings at intervals while releasing
the sex pheromone from her extruded gland. If her wings are
temporarily at rest when the male contacts her, she immediately resumes wing vibration. The male hovers in flight
below the female. Typically, he touches the gland with the
antennae, first one and then the other; then he touches the
gland with the tarsus of one prothoracic leg. These steps
are sometimes omitted and perhaps enable the male to
heighten stimulation sufficiently to release his copulatory
behavior. The male now flies to a point close beside the
female, with their vibrating wings overlapping. He curves
his abdomen toward that of the female and everts the scent
brushes (which are often stated to function as an aphrodisiac dispenser, though this has not been proved) and genital
claspers at the posterior of his abdomen. If successful
contact is made with the female, the moths move to a position with heads opposite each other, reassume a resting
position, and spermatophore transfer takes place. This
behavioral sequence is based on laboratory investigations

and has not been reported for any nocturnal moth in nature. None of the above behavioral steps require the presence of the intact female. To the nervous system of the male, the pheromone appears to represent the female. Although visual stimuli from the female body and air movement stimuli from her vibrating wings help to direct close range (centimeters) orientation and the directing of the abdomen in the copulatory attempt (Shorey and Gaston, 1969), the female can be replaced by a drop of pheromone extract on filter paper. The male orientation and copulatory behavior patterns will still be performed normally, directed at the point from which the pheromone is evaporating. The male can be likened to a computerized robot, with the repertoire of behavior programmed into its nervous system. This outline of behavioral steps illustrates an important point: a bioassay is based on the normal responses of the insect to the pheromone, and every attempt should be made to keep the unnatural environment of the laboratory from interfering with the normal behavior.

It is not necessary to observe the whole sequence of behavioral steps in a pheromone bioassay. In fact, it is advisable not to base the assay on the sequence, but instead to select only one step, such as the forward movement of the antennae, wing extension or vibration, upwind flight, or copulatory attempts. This is because different pheromone concentration thresholds are probably required for the initiation of the various steps. A very low concentration may be sufficient to induce an early step such as antennal or wing extension as compared with a later step such as copulatory attempts. This has been demonstrated experimentally by Schwinck (1955) using the silkworm moth <u>Bombyx</u> <u>mori</u> (L.) (Bombycidae) and by Traynier (1968) using the Mediterranean flour moth, <u>Ephestia</u> kuehniella (Zeller) (Phycitidae). For both species, the greater the dilution of pheromone, the earlier in the hierarchy of behavioral steps is male behavior terminated. Therefore, for assay precision, which is related to small variance and steep slope to assay line, only one or a few contiguous steps should be selected as assay indicators. To detect the lowest possible pheromone concentration, an early step should be selected.

Except for electroantennograms, which are a specialized single organ assay, all Lepidoptera female sex pheromone bioassays that have been devised utilize one of

the above-mentioned behavioral steps. Examples include
assays based on: (1) wing vibration by males of the pink
bollworm, Pectinophora gossypiella (Saunders) (Gelechiidae)
(Berger et al., 1964), and a number of species of Noctuidae
(Gaston and Shorey, 1964); (2) upwind orientation in flight
tunnels by males of the cotton leafworm, Prodenia litura
(F.) (Noctuidae) and Agrotis ypsilon Rott (Noctuidae)
(Flaschentrüger and Amin, 1950), and the Indian-meal moth,
Plodia interpunctella (Hübner) (Phycitidae) and the
Mediterranean flour moth (Schwinck, 1953); and (3) copulatory responses by tethered males of the gypsy moth,
Porthetria dispar (L.) (Lymantriidae) (Block, 1960). A
variation of the copulatory behavior that may be utilized
for assay purposes with some species is the tendency of
males in a pheromone-permeated atmosphere to circle around
with vibrating wings in a "dance", extending their clasping
organs toward the pheromone source or nearby males (Schwinck,
1955).

A qualitative bioassay is based on male behavioral
response to a single, often unknown pheromone concentration
and can only indicate whether or not the pheromone is detectable. This type of assay may be very useful for spot-
checking chemicals for biological activity. However, the
investigator must beware, because if only 1% of the pheromone remains in a sample and 99% has been destroyed or
lost, the qualitative assay can still provide the information, "Yes, pheromone is present". Then the investigator
may proceed unaware of his losses. Thus, Butenandt et al.
(1961a,b) obtained a yield of 0.024 µg of isolated sex
pheromone per female silkworm, although Steinbrecht (1964)
later showed that each female contains about 1.5 µg. For
this insect, there may have been a 98% loss of pheromone
during chemical purification. To monitor and to help guard
against such losses, quantitative assays should be used.

A quantitative bioassay can answer the question, "Relative to some standard, how much chemical is present?" In
accordance with assay principles developed for pharmacologi-
cally active chemicals, insect responses increase on an
additive basis as concentrations are increased on a multi-
plicative (logarithmic) basis. The difference between 1
and 10 µg is just as meaningful to the insect as the difference between 10 and 100 µg or 10^{-5} and 10^{-4} µg. This principle, related to the least noticeable proportional difference or the Weber fraction, has been expanded by Wright
(1964a). When responses are recorded as the percentage of

individuals responding, a probit or logit transformation of the percentages plotted against the logarithms of the concentrations tested usually yields a straight line. Such lines should be constructed for standard (known) and unknown chemicals. By comparison with the standard curve, the results of chemical purification, the activity of extracts and fractions, the activity of isolated and synthesized compounds, and reductions or increases in activity due to suspected behavior-inhibiting chemicals or pheromone potentiators can be determined quantitatively. In the absence of knowledge of absolute quantities of pheromone in the standard (weight per unit volume) arbitrary values, such as female equivalents (number of extracted female glands per test volume), can be assigned. The activity of other samples relative to these arbitrary quantitative measurements can still be determined.

A quantitative bioassay, comparing unknowns with a reference standard curve, is also the basis for physiological experiments such as quantitative comparisons of the pheromone-responsiveness of males of different ages, or at different times of the day, or under different light intensities (Shorey and Gaston, 1964).

The ultimate demand for rigidly characterized conditions in a quantitative bioassay occurs when the investigator desires to specify the behavioral activity of a compound, for example, that the threshold concentration required to induce wing vibration, antennal responses, or some other behavioral response, was X µg. There is a great deal of information available regarding the extreme biological activity of sex pheromones (Jacobson, 1964), but unfortunately almost none of it is interpretable. For instance, the amounts (in grams) of synthetic pheromone needed to induce behavioral responses in laboratory assays are stated to be about 3×10^{-20} (Butenandt and Hecker, 1961) vs. 1.5×10^{-12} (Steinbrecht, 1964) for the silkworm, 1×10^{-12} (Shorey and Gaston, 1965a) vs. 2×10^{-12} (Berger, 1966) for the cabbage looper, and 3×10^{-12} for the fall armyworm (Sekul and Sparks, 1967). But what do these values represent? They may have been based on a "just noticeable" behavioral response in some cases, and a 50% response in others. They were all based on amount of material placed on the evaporation substrates, which varied from glass to filter paper. Also they were all conducted with differing systems of air movement, usually unspecified, carrying the pheromone molecules toward the insect antennae. The amount of pheromone placed on the substrate does not reveal the concentration present in the air space around the males. In order to compare biological activity for different insects, the minimum quantitative measurement necessary is that amount of pheromone present in a given volume of air directed at the insect. Boeckh et al. (1965) calculated that 200 molecules of synthetic silkworm pheromone per ml of air would induce 50% of the males to respond (BR_{50}) within 30 seconds, and Shorey et al. (1967), estimated that 5×10^5 molecules of synthetic cabbage looper pheromone per ml of air were needed to induce a similar response. To have these figures more closely comparable, the velocity of the air passing through a given cross-sectional area in the assay chamber should also be taken into account. Based on the calculations shown in Table 1, the concentrations of pheromone needed to induce a BR_{50} are 200 molecules and 10,000 molecules per second passing through a cross-sectional area of 1 mm^2 for the silkworm and cabbage looper, respectively. Only when the results of various laboratories are stated in this way can minimum quantities of pheromone necessary to induce a certain level of biological activity be compared.

B. Potential Problems In Bioassay Design

To illustrate the need for rigorous control of conditions while conducting quantitative bioassays, the following list of common or potential problems has been constructed. Although this is a negative approach, the problems often become evident, while their solutions may not.
1. No known air flow. The pheromone is introduced into the air space surrounding the insects on a substrate (glass rod, filter paper) held close to them, or is puffed toward them from a pipette activated by a rubber bulb. Because there is no control over the flow of pheromone-permeated air with respect to the insect's location, and because of incomplete mixing of pheromone-permeated air with air in the test chamber, results may not be reproducible.
2. Release of a nonreproducible quantity of pheromone from pipette. In the commonly employed technique of drawing a pheromone solution into a pipette, allowing the solvent to evaporate, and then forcing air through the pipette toward the test insects, the number of pheromone molecules in the

Table 1

Calculations for comparing sex pheromone quantities required for production of a BR_{50} (50% behavioral response) by males of the silkworm (Boeckh
et al., 1965) and cabbage looper (Shorey et al.,
1967).

	Species	
	Silkworm	Cabbage looper
Pheromone (g) on evapora-tion source	$1x10^{-10}$	$1x10^{-11}$
Molecules/sec. leaving source	$7x10^6$	$4x10^7$
Air volume (cm^3/sec.)	$4x10^4$	83
Molecules/cm^3	$2x10^2$	$5x10^5$
Cross-sectional area of stimulating air current at antennae (mm^2)	$4x10^4$	$5x10^3$
Air volume (cm^3) pass-ing through cross-sec-tional area of 1 mm^2/sec.	1.0	$17x10^{-3}$
Molecules passing through cross-section-al area of 1 mm^2/sec.	$2x10^2$	$1x10^4$

air will depend on the time interval since the rubber bulb was last squeezed. This happens because, in the absence of air movement, the pheromone is continuously evaporating from the walls into the lumen of the pipette.

3. Release of nonreproducible quantity of pheromone from filter paper. The cabbage looper sex pheromone evaporates from filter paper at a rate of about 8% per minute (Shorey and Gaston, 1965a). In the minute following impregnation of 1 μg on filter paper, about 0.08 μg will evaporate into the air stream. Five minutes later, only about 0.05 μg will evaporate in one minute. Therefore, reuse of treated papers will bias quantitative results.

4. Reuse of assay males. Responsiveness of male Lepidoptera is generally inhibited after a single exposure to sex pheromone (Shorey et al., 1968). It may not be advisable to use males for assay more than once per day.

5. Prior to the assay, males not isolated from air coming into contact with female moths. The same effect is created as is discussed in No. 4.

6. No selection of a single behavioral response as the assay indicator. This arises from lack of appreciation that greatly different pheromone threshold concentrations may be required to induce the various types of male responses (Section II, A).

7. Improper attention to the fact that physiological variables (male age, circadian rhythms, pheromone exposure history) may cause pheromone thresholds to vary as much as 1000-fold (Shorey et al., 1968).

8. Improper attention to environmental variables (light intensity and quality, temperature), which may cause the same effect as No. 7 (Shorey et al., 1968).

9. Nonspecificity of assay behavior. If the given behavior (wing vibration, copulatory attempts) can be released by other chemicals as well as the sex pheromone, adequate controls must be run using solvents and nonpheromone extracts to assure that the sex pheromone is the chemical producing the observed effect (Block, 1960).

10. Competing stimuli. This may be particularly important in flight tunnels, in which pheromone-stimulated males may orient toward a visible light source rather than toward the pheromone source (Shorey and Gaston, 1965b).

11. Contagiousness of response. With more than one assay insect in the test chamber, the movement of the first responding insect may cause the others to move (respond) also.

12. Impure air supply. The air entering the assay system may have to be "scrubbed" to remove oil, water, and other impurities that could affect the responsiveness of the test males.
13. Unknown seasonal factors. Colonies of noctuid moths maintained at Riverside, California, vary in vigor and in pheromone-responsiveness depending on the season of the year. These variations make it necessary always to compare experimental results with concurrently conducted standard curves.
14. Colony breeding history. The same considerations apply here as in No. 13.

III. Orientation Of Males To Pheromone-Releasing Females

Although this review is written from the point of view of nocturnal Lepidoptera, many of the orientation mechanisms to be discussed pertain to other insect groups also.

Before male orientation can occur, the sex pheromone must be dispersed from the vicinity of the female, and she plays an active role in its dispersal. The lepidopteran sex pheromone gland is located typically in the intersegmental membrane between the eighth and ninth abdominal segments (Section IV, C). The gland form and position in this area vary from species to species. When the female is at rest, the gland is not normally exposed to the air, because the ninth and eighth segments are retracted into the seventh segment. Also, many types of glands are invaginated into the body cavity. At the appropriate time for mating, the female can actively extrude the terminal abdominal segments and the glandular area. Females of many species alternately protrude and retract the sex pheromone gland (Götz, 1951; Doane, 1968; Kettlewell, 1946). This may have the effect of coding the signal by producing discrete pheromone pulses of high and low concentration in the air moving from the female. Females of some species vibrate their wings while releasing pheromone; this could constitute another coding, at wing-beat frequency. However, so far there has been no demonstration of an influence of a pulsing release of pheromone on the orientation responses by male moths.

The female apparently releases the pheromone from the gland surface by evaporation and fans the molecules away posteriorly with her vibrating wings. Then the pheromone is subject to the mixing and movement processes of the air.

A. Diffusion Gradients

    In an environment having no air movement, pheromone
molecules evaporating from a gland would diffuse outward in
all directions in an expanding sphere, with a gradient of
concentrations from most dense at the center to least dense
at the periphery. An orientation mechanism, based on the
responding organism being able to sense the directional
component of a uniform concentration gradient and to steer
positively with respect to it, is the only true chemotaxis.
However, even if such a gradient could be maintained for
any distance in air, it appears unlikely that an insect
possesses the sensory apparatus to allow detection of the
direction of the gradient over a distance of more than a
few centimeters from the source (Dethier, 1957; Neuhaus,
1965; Shorey and Gaston, 1967; Wright, 1958). It is equally
unlikely that a concentration gradient could remain intact
in open air over a distance of more than a few centimeters
(Wright, 1958; Bossert and Wilson, 1963).
    Air is constantly in motion, if not because of wind
currents, then because of convection currents caused by the
uneven heating of different segments of air. The air does
not flow in a smooth laminar nature, but is characterized
by disruptive, turbulent, shearing flows on both a micro-
and macro-scale.
    As a result of the above arguments, some investigators
such as Fabre (Teale, 1961), Laithewaite (1960), and
Callahan (1965), have postulated that infrared or other
electromagnetic energy radiating from the female must serve
to orient the males. However, male orientation to a pheromone source, apparently guided only by the presence of pheromone molecules in the air and the direction of air movement, has been demonstrated for a number of species of
Lepidoptera (Shorey et al., 1968). The question is not,
"Do moths orient from a distance toward a pheromone source?"
It is, rather, "How do they do it?"

B. Anemotaxis

    Because the air mass as a whole is usually moving in
some average (prevailing) direction, the cloud of pheromone
molecules will extend away in that direction from the female.

Pheromone molecular dispersion in the moving air is largely
a function of the mixing due to air turbulence. On the
average, the pheromone concentration will still be highest
in the vicinity of the female, and the zone bounded by any
given pheromone concentration will assume a long ellipsoidal shape (Wright, 1958; Bossert and Wilson, 1963). The
physical dimensions of the ellipsoidal shape will depend on
the evaporation rate of pheromone from the female, the
average wind speed, and other environmental factors. It is
apparent, however, that the ellipsoid is only an average of
many hundreds of individual cases; at any given moment the
different pheromone-laden densities of air will shear and
reverse and branch off in filaments according to micro-
environmental air currents.

Among the many observations of upwind orientation toward a pheromone source, that of Schwinck (1958) is usually
cited as the classical example. She observed that when
males of the silkworm running free on a table top were
stimulated by the female sex pheromone, they oriented accurately toward the source of an air stream even though the
females were removed from the source. Her proposal that
the males do not orient (steer) with respect to the source
of the pheromone, but that the pheromone stimulation causes
them to orient anemotactically with respect to the direction
of air flow, has received considerable experimental support.
It is accepted, with reservations, by many investigators of
olfactory behavior.

A major question raised is, "How do flying insects sense
the direction of the moving air?" This has been clarified
for certain day-flying insects, such as mosquitoes (Kennedy,
1939) and Drosophila (Kellog et al., 1962), which have been
found to use visual contact with the ground to stabilize
upwind orientation. If the apparent movement of the ground
over which the insect passes is parallel to the long axis
of the insect's body, then it is flying either up- or downwind. Cross-wind flight will cause drift of the insect
with respect to the ground and a resulting apparent angular
movement of the ground with respect to the long axis of the
insect.

Some light is always available at night, and it is
reasonable to suspect that the darkness-adapted eyes of a
male moth have sufficient acuity to enable a similar visually
based navigation. The male moth might also possess the
necessary sensory mechanisms, based on differing inertias

of appendages relative to the whole body, to permit perception of the relative directions of air movement at the boundaries between the various shearing air currents through
which it flies.

C. Orientation by Odor Pulses

     Wright (1958) focused attention on the fact that mole-
cular density within an odor cloud is not uniform. He felt
that it was impossible to assume that an insect could back-
track along the intricate and rapidly shifting odor trails
within the cloud, and proposed an orientation mechanism
based on detection by the insect of the changes (pulses) in
pheromone concentration as it flies through the cloud. If
an insect flies toward the pheromone source, it encounters
pulses at an increasing frequency. If it flies across
the cloud or down the cloud away from the source, it encounters pulses at a constant (on the average) or decreasing
frequency. Wright proposed that if the insect maintained a
fixed flight path when encountering pulses at an increasing
frequency and a short zig-zag path when encountering pulses
at a constant or decreasing frequency, it may reach the
source. Although these ideas have received no experimental
support (Kellogg et al., 1962), they are of value in that
they emphasize the disrupted nature of the odor cloud
through which a male moth must fly.

D. Behavior at Odor Boundaries

     The behavior of flying mosquitoes (Daykin et al.,
1965) and Drosophila (Kellogg et al., 1962) at boundaries
between favorable (i.e., containing a stimulating odor) and
neutral air streams was studied. The insects tended not to
turn when entering the favorable air stream, or, if turning, they oriented upwind. They tended to turn back when
leaving the favorable air stream. The turn by Drosophila,
within 0.1 sec after leaving the favorable stream, causes
the insect to fly at roughly a right angle to the wind. If
it does not regain the scent, the insect may fly across the
wind in the reverse direction or upwards or downwards,
still at a right angle to the wind. These cross-wind
"casts" last about 0.3 sec. If the favorable air stream is
still not located, it then flies downwind for a distance of
6 to 24 inches before resuming the cross-wind flights.

It appears likely, from the research of Traynier (1968) with the Mediterranean flour moth, that a similar mechanism might control the orientation of male Lepidoptera at boundaries between stimulating (i.e., pheromone concentration above the behavioral threshold) and neutral (i.e., concentration below the threshold) air streams. A moth flying within the filamentous, disrupted odor cloud is apt to leave a favorable air stream many times. If we assume that a turn at a right angle to the micro-wind direction is made every time the male moves into an area of lesser pheromone concentration and that an upwind turn is made when it enters an area of higher concentration, there is a fair likelihood that it may reach the female.

E. Adaptation and Thresholds for Stimulation

There must be some minimum density of pheromone molecules in the air that will trigger upwind orientation by the receiving male. Wright (1965) suggests that this density may be about 10 molecules/mm^3. Experimental data (Section II, A, Table 1) indicate that molecular densities for a 50% activation of males of the silkworm and the cabbage looper are about 0.2 and 500 molecules/mm^3, respectively. It is not known whether these molecular densities also represent the thresholds for anemotactic behavior.

If an above threshold pheromone concentration in air were distributed uniformly, the male would probably soon stop responding (Shorey et al., 1967). This is because antennal sensory cell adaptation and central nervous system habituation would soon convey the message that no pheromone was present when, in fact, a uniform concentration was present. A male flying from an area containing a high pheromone concentration into an area of lesser concentration will receive a similar message; with the male conditioned to the high concentration, the lower concentration is the behavioral equivalent of zero pheromone. Flying through the disrupted pulses in an odor cloud may be advantageous to the orienting male. Although the lower concentration is nonperceptible, it will permit the level of sensory adaptation to lessen so that the male can again respond positively when it encounters a new area of increased concentration.

F. Synthesis of Hypotheses

The factors discussed above are combined in the follow-

ing proposed mechanism for orientation of a male moth toward a pheromone source. The male may already be flying
when it enters an area containing a pheromone concentration
above its behavioral threshold, or an above-threshold concentration in the air stream flowing over a resting male
may stimulate it to fly. The only influence of the direction of the macro-air flow (prevailing wind direction) is
to move the total mass of pheromone-laden air in a given
direction from the female. The male senses and responds
only to the micro-air flow, which may sometimes be in a
direction opposite to that of the macro-air flow. Every
time the male enters an area of higher effective concentra-
tion--one that is above the threshold concentration level
to which the male has adapted and which thus stimulates the
male--it orients upwind in the micro-air flow. Every time
the male enters an area of lower concentration--one that is
below the threshold to which it has adapted--it steers in a
programmed way, perhaps at right angles to the wind, with
the direction of the turn perhaps opposite to that of its
previous turn. The program for this steering behavior that
occurs when the male has lost odor stimulation is certainly
not random; evolutionary selection pressures will have
favored development of behavior patterns that maximize the
likelihood of the male regaining odor stimulation. The
state of adaptation to high concentrations that the male
has encountered will lessen each time it enters an area of
lower concentration, so that re-entry into a new area of
high concentration will again provide nervous stimulation.

G. Effect of Other Aggregation Stimuli

The efficiency of the orientation mechanism that causes
a male to locate a female may be aided by the influence of
environmental aggregation stimuli, especially those coming
from host plants. For instance, in some moth species males
and females may tend to congregate in the canopy of trees
of certain species. The canopy may set the outside limit
of the universe within which most males fly. Within the
canopy, a number of females may release pheromone with the
effective zones (above male behavioral threshold) from
several females overlapping. This overlap presents no com-
plication for male orientation. The male still will only
respond to stimuli indicating the changing pheromone concentrations and micro-wind directions that it encounters.

Due to the interweaving of the scent clouds, the male may not necessarily end up mating with the same female whose pheromone molecules he first encountered.

H. Short Range Orientation

When does the general upwind orientation of pheromone-stimulated males cease? In the vicinity of the females, a variety of stimuli have been demonstrated to induce short range orientation and copulatory behavior. Males of the cabbage looper (Shorey and Gaston, 1969), the Mediterranean flour moth (Traynier, 1968) and the gypsy moth (Doane, 1968) utilize visual stimuli from the female body but only when in the presence of the pheromone. However, even in the absence of the female a high concentration of pheromone evaporating from filter paper is sufficient to cause the males to orient to the paper and to cease upwind orientation. It is possible that a sharp differential between very high and very low pheromone concentrations over a short distance, plus a nearby solid substrate, conveys the behavioral message of "female nearby" and causes the initiation of short range orientation behavior, which is not necessarily guided by air movement.

I. Effective Distances for Orientation

None of the great distances for male orientation to pheromones that are cited in the literature are subject to experimental proof. They are based on arrival of males, usually marked, from known distances downwind from the pheromone source. There is no way to determine how far these males flew without pheromone stimulation before they encountered a concentration above the behavioral threshold.
An average answer to the question of orientation distances can be approximated by experimental determination of the following factors:
1. Female pheromone release rate per unit of time. This has not been determined for any insect species, though it probably falls between 10^{-9} and 10^{-7} g/min for the silkworm and the cabbage looper (Steinbrecht, 1964; Shorey and Gaston, 1965a).
2. Dispersal characteristics of the pheromone molecules in the air flow of interest, giving average molecular densities at various points in space relative to the releasing female.

Formulae are available for this calculation, but the value obtained is only an average and does not refer to any specific case (Section III, B).

3. Threshold concentrations for responsiveness of the receiving male. The concentrations of pheromone per unit volume of air required to induce BR_{50} wing vibration responses by males of the silkworm and the cabbage looper have been determined (Section II, A, Table 1); however, the concentration needed to induce upwind orientation may or may not be higher.

IV. Chemical Evolution and Species Specificity

It is difficult to define species specificity with regard to sex pheromones. Despite the word specificity, which indicates a strict separation between each species, one must think in terms of the relative degree of species specificity and ask to what extent it occurs.

By nonspecies specificity of a lepidopteran sex pheromone, we mean that the males of species B respond behaviorally to the pheromone emitted by females of species A. There are at least three mechanisms that could account for this nonspecificity: (1) Females of each species may use the identical chemical as their sex pheromone; males of each species respond maximally to this chemical and will therefore approach females of either species. (2) Females of species A use chemical A as their pheromone, but also release chemical B from their gland. Males of species A respond to chemical A, but do not necessarily respond to chemical B. Females of species B release chemical B from their gland, and the males of species B respond to chemical B. They will therefore respond to the pheromone mixture emitted from females of species A, as well as to that from their own females. (3) Females of species A use chemical A as their pheromone and do not produce chemical B. Females of species B use chemical B and do not produce chemical A. Chemicals A and B are sufficiently similar that males of species B, although responding maximally to chemical B, will also respond (perhaps with a much higher threshold) to chemical A.

It may be difficult to determine from observations of nonspecificity of sex pheromone behavior which of the above three mechanisms is operating. The literature contains

numerous examples of nonspecies specificity of insect sex
pheromones (Schneider, 1965; Shorey et al., 1968). The list
includes many cases of nonspecificity among species that
are currently classified in different genera and a few cases
of nonfamily specificity. Of special interest is the observation of Schneider (1962) that antennae of males of the
silkworm responded weakly to stimulation with the sex pheromones of several species of Saturniidae, but not vice versa.
Silkworm males responded behaviorally to the saturniid
pheromones as well, and the response was much less intense
than was that to their own sex pheromone. This would indicate that mechanism No. 2 or 3 is causing a nonspecies
specificity. On the other hand, Shorey et al. (1965b)
could find no quantitative behavioral or chemical (GLC) evidence indicating that the sex pheromones of the cabbage
looper and the alfalfa looper, Autographa californica
(Speyer) (Noctuidae) were not identical. Mechanism No. 1
may be operating here to cause nonspecies specificity.

A. Reproductive Isolation

A brief description of a species is that it is an
interbreeding unit reproductively isolated from similar
units with which it may coexist. Mayr (1963) has separated
reproductive isolating mechanisms into two categories, those
that prevent interspecific matings (premating mechanisms)
and those that reduce the success of interspecific matings
(postmating mechanisms). Postmating mechanisms operate
when a male of one species successfully transfers sperm to
a female of another species, but vigorous breeding offspring are not produced because of physiological or genetic
incompatibilities between the two species. Postmating
mechanisms involve wastage of gametes and are therefore
less efficient than premating mechanisms. Premating isolation mechanisms, which prevent wastage of gametes, are
highly susceptible to improvement by natural selection.
Premating mechanisms can be subdivided according to mechanical, spatial and temporal, or behavioral isolation.
Mechanical isolation occurs if the genitalia of species
B are sufficiently different from those of species A that
males of B cannot effectively copulate with females of A.
Males of either of the related noctuid species, the corn
earworm, Heliothis zea (Boddie), and the tobacco budworm,
H. virescens (F.), responded to the female sex pheromone
of the other (Shorey et al., 1965). However, all males of

267

the corn earworm that succeeded in coupling with tobacco budworm females could neither transfer spermatophores nor separate from the females. This is an extreme case of mechanical reproductive isolation in which not only was gene flow between the two species prevented, but also those males and females that cross-coupled did not survive. These observations were based on laboratory behavior, but other isolating mechanisms may keep the two species from coming in contact in nature.

Spatial or temporal isolation can be achieved by the two species having different host specificities, aggregation stimuli, times of day at which mating occurs, or seasonal or geographic distributions (Alexander, 1964).

Behavioral isolation can operate if close range courtship stimuli, that must be exchanged between males and females before mating can occur, are not matched between the two species, even though the males may initially approach the wrong females because of nonspecific sex pheromones. The ultimate efficiency in behavioral isolation for Lepidoptera is a complete specificity of sex pheromones, so that from a sexual point of view neither species knows that the other one exists. However, it is evident that one does not have to assume that two species must have effectively different sex pheromones in order to be reproductively isolated from each other.

B. Sex Pheromone Evolution, Early Stages

Moore (1967) has made the following comments concerning the possible evolutionary history of insect sex pheromones. Early primitive insects were cryptic and restricted in their habitat, largely by a need for moist conditions. The sense of smell was not developed in these insects, and food and mates were located by random searching. With the development of a more specialized cuticle and the evolution of wings, insects were released from their restricted habitat. Becoming more mobile, they were better able to adapt to a wide range of discrete habitats. Their feeding habits became more specific, and some degree of olfactory ability was required to detect and discriminate between food sources. Subsequent evolution in this direction, with a continual sharpening of olfactory powers, preadapted the insects to a chemical means of communication. It is assumed that at this evolutionary stage the activities of one sex may have

rendered the food more attractive to the other sex through
the production of metabolites. This attribute would be
selected as an advantageous biological characteristic, increasing the likelihood of bisexual aggregations and mating.
Further evolution might have led in some cases to the combination of the several signal components into a single
highly specific chemical structure. Now the species could
well have become independent of the food lure as an aggregation stimulus for mating, and further mutual sharpening
of both scent and receptor mechanisms could have led to the
specifically tailored sex pheromone molecules characteristic
of many Lepidoptera today.

These speculations of Moore provide an interesting
reflection on the possible early evolutionary history of
insect sex pheromones. However, they do not consider the
question of how the great diversity of sex pheromones used
by present day Lepidoptera arose.

C. Sex Pheromone Evolution in the Lepidoptera

We will start with the assumption that sex pheromone
communication is a very primitive characteristic of Lepidoptera, and that it has evolved from some common ancestral
type to the present day variety of glandular structures,
chemical structures, and associated behavior patterns. This
assumption is based on morphological and chemical evidence.

The morphological evidence is based on the location of
sex pheromone glands (Götz, 1951; Jacobson, 1965; Shorey
et al., 1968). The glands are always located on the posterior portion of the female abdomen, typically in the
intersegmental membrane between the eighth and ninth abdominal segments, though occasionally between the seventh and
eighth segments. Although the gland has evolved into a
variety of shapes and positions within this location, the
common location is taken as a primitive characteristic.

Chemical evidence is based on the similarity in structure of seven Lepidoptera sex pheromones.

Superfamily Bombycoidea
The silkworm: trans-10-cis-12-hexadecadienol (I)
(Butenandt et al., 1959, 1961a, b).

$$CH_3(CH_2)_2CH=CHCH=CH(CH_2)_8CH_2OH$$

(I)

269

H. H. SHOREY

Superfamily Gelechioidea
    The pink bollworm: 10-propyl-trans-5,9-tridecadienyl
acetate (II) (Jones et al., 1966)

$$[CH_3(CH_2)_2]_2C=CH(CH_2)_2CH=CH(CH_2)_3CH_2OCCH_3$$
$$O$$

(II)

Superfamily Noctuoidea
    The gypsy moth: 10-acetoxy-cis-7-hexadecenol (III)
(Jacobson et al., 1961).

$$CH_3(CH_2)_5CHCH_2CH=CH(CH_2)_5CH_2OH$$
$$OCCH_3$$
$$O$$

(III)

    The cabbage looper: cis-7-dodecenyl acetate (IV)
(Berger, 1966).

$$CH_3(CH_2)_3CH=CH(CH_2)_5CH_2OCCH_3$$
$$O$$

(IV)

    The fall armyworm: cis-9-tetradecenyl acetate (V)
(Sekul and Sparks, 1967).

$$CH_3(CH_2)_3CH=CH(CH_2)_7CH_2OCCH_3$$
$$O$$

(V)

Superfamily Tortricoidea
    Argyroploce leucotreta Meyr. (Eucosmidae): trans-7-
dodecenyl acetate (VI) (Read et al., 1968).

CONTROL OF INSECT BEHAVIOR

$$CH_3(CH_2)_3CH=CH(CH_2)_5CH_2OCCH_3$$
$$O$$

(VI)

The red-banded leaf roller, Argyrotaenia velutinana (Walker) (Tortricidae): cis-11-tetradecenyl acetate (VII) (Roelofs and Arn, 1968).

$$CH_3CH_2CH=CH(CH_2)_9CH_2OCCH_3$$
$$O$$

(VII)

The validity of structures (II) and (III) as sex pheromones of the given species has been questioned by Eiter et al. (1967). They synthesized the chemicals and reported that they caused no behavioral responses by males of the respective species. For the purpose of the present discussion, it is assumed that if the original identifications were incorrect, the correct formulas are apt to be similar to those listed above.

These sex pheromones are all unsaturated primary alcohols or their derivatives and thus show an intriguing similarity in structure. As the moths themselves are not closely related, but belong to four different superfamilies, it is tempting to infer, as did Moore (1967), that this type of structure is general throughout the Lepidoptera. A further inference based on their similarity is that this type of chemical structure is derived from some common, primitive type.

But how can we account for the evolution of the array of highly specific chemicals in present day Lepidoptera? For the few synthetic sex pheromones and similar compounds that have been altered slightly, it is apparent that the males have a very narrow band of discrimination (Berger and Canerday, 1968; Jacobson, 1965; Jacobson et al., 1968; Shorey et al., 1968). Peak male responsiveness occurs to the pheromone itself, and the minor alterations cause responsiveness to decrease greatly. However, the major point to be made is that males will respond, albeit on a much less sensitive level, to chemicals modified somewhat from the exact sex pheromone structure.

271

Assume that the female does not release only one chemical, the sex pheromone, from her gland. It seems highly unlikely that the biosynthesis procedure is sufficiently exact that no other chemical besides the sex pheromone is produced. In fact, through inexactness of synthesis, and a variable genetic makeup that may cause the gland to make other chemicals, there may be dozens of related chemicals produced in and released from the gland. Of course, the major production will be the sex pheromone, since individual females that minimize release of the correct pheromone will be discriminated against in sexual selection.

This aspect of a female moth releasing a number of chemicals is similar to the thesis of Wright (1964b); but he further proposes that a medley of chemicals, not one specific chemical, functions as the sex pheromone. I differ from that assumption here by proposing that although a female moth may release a medley of chemicals, only one of them has been selected behaviorally as the sex pheromone.

Let us assume that through geographic isolation species A was divided into two non-interbreeding populations, A and A'. Population A' has adapted to its new environment and has evolved into species B. While in its new environment, through genetic reconstruction, the quantitative and qualitative composition of minor chemicals produced in the sex pheromone gland of B may shift from the composition in A. After species B has become re-established in the same area with species A, we find that those individuals that cross-mate do not produce viable offspring. There will now be strong selection pressure in favor of any behavioral divergence between the two species that will prevent attempts at cross-mating and thus will preserve gametes within one species. Any inherent tendency, although it may be slight, for a male to recognize and respond to one of the minor chemicals released from the sex pheromone gland of its own species, but not from that of the other species, would be rapidly incorporated into the species behavior by natural selection and could lead to one of the species now having a sex pheromone distinct from that of the other. Since any male that still would respond to the wrong pheromone would not tend to contribute to its own species gene pool (wastage of gametes), perception by males of species A of its own sex pheromone, but not that of species B, and vice versa, would become sharply defined.

It should be stressed again that the above system

proposes a method for production of behavioral reproductive isolation through the mechanism of different sex pheromones in the two species. Instead of selection favoring development of different sex pheromones to prevent gamete wastage, it may in some species favor other premating isolation mechanisms.

V. Potentials For Behavioral Control Of Pest Species

Many of our most injurious agricultural and forest pests are larvae of nocturnal Lepidoptera. Control of Lepidopterous larvae could be achieved by keeping males and females from coming together for mating.

In the communication sequence that leads to mating, the female releases her sex pheromone into the air (Section II, A). It is this chemical message that activates the male at some distance downwind from the female. Perceptually, in distance communication, the pheromone is the female (Shorey et al., 1968). No other factor is needed to initiate the sequence of behavioral steps that causes the male to approach her vicinity. Other messages, such as visual stimuli released by the female, may play a role in short range (centimeters) communication (Section III, H). However, even at a short range, without pheromone stimulation the male's "female-searching" and copulatory instincts may not be activated. This has been demonstrated in the laboratory by confining sexually mature males and females of the cabbage looper together in glass jars. When the air in the jars was saturated with synthetic sex pheromone, no mating occurred (Shorey et al., 1967). At the appropriate time of day, the females assumed the characteristic pheromone-releasing posture with the pheromone gland extruded and the wings vibrating. The spatial characteristics of the jars prevented any male from being further than 15 cm from a female. The lack of response by the males probably indicated that the pheromone evaporating from the females was indistinguishable to them from the high ambient concentration in the jars.

This necessity for airborne chemical communication between males and females may be a very vulnerable point in the life cycle of moths. The question is, "How can we manipulate the sex pheromone of a particular moth species to ensure that the two sexes do not come together for mating?"

First, we must consider how efficient the control
should be, "Must we effectively prevent 50%, 90% or 99+%
of the males from reaching the females?" It appears characteristic of many Lepidoptera that the male:female sex
ratio is about 50:50. In the case of the cabbage looper and
probably most other species, the males are multiple maters
and can inseminate one female per night during most of their
adult life (Shorey, 1966). Although the average life span
of a male cabbage looper in the field is subject to a number of environmental factors, especially temperature, 14
days is probably a reasonable estimate. During that time,
the male may mate 10 times, probably with 10 different
females. We could reduce the effective male:female ratio
to 10:90 and most females would probably still be insemi-
nated. If we achieved a 90% control of female egg production by reducing the ratio to 1:99, this would often not
be considered as an economic control of the pest population.
Therefore, over 99% of the males would probably have to be
prevented from finding females before the behavioral control
program could be efficient. The control of males must be
essentially complete.

A. Methods for Behavioral Control

Although there are many possible variants of technique,
the proposed methods for population control of moths using
sex pheromones can be divided into two categories.
1. The Use of Pheromone-baited Traps.
This method preys on the normal instinct of sexually
responsive males to orient toward a pheromone source. The
males may be killed at the trap, or they may be sterilized
there and be returned to the field to mate with and effectively sterilize wild females. On the basis of theoretical
models, Knipling and McGuire (1966) calculated that sterili-
zing the males would provide only a slight advantage in
population control over killing them at the traps. In the
trapping method, the goal is to out-compete the females at
their own game--to get 99% of the males to the traps before
they get to the females.
2. The Use of Pheromone-evaporating Substrates to Permeate
the Air.
The air must be permeated to a sufficiently high level
that the relatively small additional increment of pheromone released by females will be imperceptible to the males,

and they will not be able to orient to females for mating. The principle here is that instead of stimulating and misdirecting male behavior, we are attempting to disrupt normal male behavior. The feasibility of this male inhibition technique has been recognized by several authors (Babson, 1963; Beroza and Jacobson, 1963; Wright, 1965; Shorey and Gaston, 1967; Shorey et al., 1968). Wright (1965) speculated as to the pheromone concentrations that might be needed. He suggested that the lowest concentration of an odorous chemical that could elicit behavior in an insect is about 10 molecules/mm^3 of air (Section III, E), and predicted that the concentration required to "saturate" the chemical receptors and thus to block the expected behavior is 10^5 molecules/mm^3 higher than the behavioral threshold. If this were the case, for a chemical having a molecular weight of about 200, the air would have to be permeated with 10^{12} molecules (about 10^{-9} g) of pheromone/liter during the time of communication for mating.

B. Case Histories

No successful control of a breeding population of Lepidoptera by the use of sex pheromones has yet been accomplished. However, the proposed procedures have been investigated on large and small scales, with varying degrees of success. The following discussion of case histories is presented as a format for analyzing requirements for a successful behavioral control program. The first two case histories involve the male trapping technique, and the second two involve the male inhibition technique.
1. Male trapping.
    a. Nun moth, Porthetria (Lymantria) monacha (L.)
    (Lymantriidae).
    Jacobson (1965) has reviewed the attempts of several European investigators during the 1930's to trap sufficient nun moth males so that most females would remain unmated and subsequent larval populations would remain below damaging levels. The techniques were based on exposing caged virgin females surrounded by sticky material on paper strips, cardboard panels, or tree trunks. Experimental plot sizes ranged to the thousands of hectares, containing thousands of traps. Although large numbers of males were attracted by the female sex pheromone and trapped on the sticky surfaces, in no case was a reduction in subsequent populations reported.

275

If we assume that the trap design was efficient, with all attracted males being ensnared, these tests probably failed because the females at the traps could not out-compete females in nature. In fact, except when the moth population density is very low (Knipling and McGuire, 1966), it would appear highly unlikely that a behavioral control program using living females as the pheromone source at the trap could be economically feasible. For instance, in the test reported by Ambros (1938), about 20,000 females used to bait traps in an area of about 2,000 hectares caused 150,000 males to be caught. But, if we assume that the sex ratio in this species is 1:1, there must have been at least 150,000 wild females within the effective radius of the traps competing with the 20,000 females in the traps. For a 99% control of males, it may be necessary to uniformly distribute 99 pheromone-releasing females in the traps for every one pheromone-releasing female in nature.

b. Pink bollworm.

Graham et al. (1966) conducted a field experiment in which approximately 2.5 traps per hectare were baited with extracted female sex pheromone and placed in a cotton field. Each trap contained a pheromone dispenser (filter paper) impregnated with 50 female equivalents of pheromone. The dispensers were replaced twice weekly. The trapping area comprised less than 4 hectares and was adjacent to a larger control area of cotton containing no traps. Despite a large capture of males at the traps, no reduction in the number of mated females or in the subsequent larval infestation was noted in the trapping vs. the control area.

This experiment yields no new information because of the lack of isolation of the trapping field from adjacent sources of moth infestation. At most, mated female moths had to fly 600 feet from the control area to be at the furthest corner of the trapping area. Captures of males in the trapping area could have no influence on the presence or the egg laying of those immigrant mated females.

c. Discussion of male trapping technique.

The sex pheromone source at the trap may consist of living females, extracts of females, or synthetic pheromone. The living female, although expensive and inconvenient to culture and handle, is in some ways a more efficient source for attracting males than extracted or synthetic pheromone. This was seen by Butt and Hathaway (1966) with the codling moth, Carpocapsa pomonella (L.) (Olethreutidae), and by

Wolf et al. (1967) with the cabbage looper. The living female has several advantages. One is timing: she releases pheromone only during the normal time of communication. Of course, it would be an even greater advantage if the lure female started to release pheromone before the normal time. Another advantage is quantity. If, as is suspected for the silkworm and the cabbage looper, the female releases and resynthesizes the equivalent of her total pheromone content every few minutes (Shorey et al., 1968), she may produce and release 100 times her equilibrium gland content in one week. Therefore, for species similar to these, an extract or a synthetic pheromone-baited trap may have to release at least 100 female equivalents of pheromone per week to equal the output of one female. However, most extract or synthetic pheromone-baited traps that are used release pheromone throughout the day, with peak evaporation occurring during the warm day-light hours when males of most moth species are not responsive. This means that the trap that loses pheromone 24 hours per day may have to release 1000 female equivalents per week to be evenly competitive during the normal mating time with one female (the female being either a pheromone source in another trap or a competitive wild lure for males in the field).

Three other factors may make the use of extract or synthetic pheromone-baited traps difficult: improper pheromone concentration at the trap orifice, the nonuniform evaporation characteristics of most pheromone dispensers, and contamination which inhibits male activity.

It is sometimes noted that if a large quantity of sex pheromone is placed on a volatilizing surface in a trap, captures of males remain low for a number of days and then increase. This could be due to the presence of a nonpersistent chemical that in some way inhibits male response to the pheromone, resulting in a low catch. After the degradation or volatilization of the inhibitor, assuming that the pheromone is still being released at an adequate rate, males respond and catches increase. A nonpersistent inhibitor of this type was found by Beroza (1967) in extracts of the gypsy moth sex pheromone.

In the absence of experiments with combined extract fractions, such as those conducted by Beroza (1967), one should be cautious in claiming that a low initial male catch at an extract-baited trap is caused by the presence of masking agents or inhibitors in the extract. Another reason for low initial catches may occur when the pheromone release rate from the trap is initially too high; when sufficient pheromone has evaporated from the source so that the release rate is reduced to a biologically suitable level, males may then enter the trap. This phenomenon has been demonstrated with the synthetic cabbage looper sex pheromone. Gaston et al. (1969) set up traps baited with synthetic pheromone covering the bottoms of reservoirs, to give a relatively constant release rate. Optimal captures occurred when the release rate was about 0.3 µg/min, while tripling the release rate to about 1 µg/min caused captures to decrease by almost 50%.

These reduced captures due to too high a pheromone concentration are probably related to the design of the trap. As suggested earlier (Section III, H), when the male encounters some high concentration of pheromone in air, upwind flight behavior may become switched off, and the behavioral program of short range searching and copulatory behavior may be initiated. This assumption is supported by observations with the cabbage looper; if males approach a trap from which a high concentration of pheromone is evaporating, they often do not enter the trap orifice. Instead, they approach and attempt copulation with the lip of the trap and then fly away. Because the pheromone concentration may have to be so carefully regulated, control of a lepidopterous species by attracting most of the males to traps may not be possible. If the behavior of the males dictates that the pheromone concentration leaving the catching portion of the trap cannot be much higher than that concentration which would be encountered in the vicinity of a living female, the traps may never be able to out-compete living females. If the concentration leaving the trap is adjusted higher, so that in most areas it does out-compete living females, trapping efficiency may be reduced and the investigator may be approaching the conditions needed for population control by male inhibition.

The increased tendency of pheromone-stimulated male moths to approach light and enter light traps may be a way out of this trapping dilemma (Shorey and Gaston, 1967). However, as shown by Wolf et al. (1967), cabbage looper males were inhibited from approaching black light traps when the concentration of synthetic sex pheromone evaporating from a source near the traps was too high.

CONTROL OF INSECT BEHAVIOR

2. Male inhibition.
   a. Gypsy moth. Gyplure (VIII), 12-acetoxy-cis-9-octadecenol.

$$CH_3(CH_2)_5\underset{\underset{O}{\overset{\displaystyle OCCH_3}{||}}}{CH}CH_2CH=CH(CH_2)_7CH_2OH$$

$$\text{(VIII)}$$

This homolog of the natural gypsy moth sex pheromone (III), is reported to be 100-fold less active than the natural material for attracting males in the field (Jacobson and Jones, 1962). Burgess (1964) reported that gyplure was broadcast from aircraft, in both liquid and granular formulations, over an island infested with the gypsy moth. The purpose of the trial was to permeate the air with sufficient gyplure that males would be unable to recognize and respond to the natural sex pheromone released from females. This required the assumption that adaptation of males to gyplure would automatically mean that they were also adapted to the natural pheromone. The experiment failed to prevent wild females on the island from attracting males for mating. Various reasons for this failure have been proposed by Waters and Jacobson (1965) and Shorey et al. (1968). Eiter et al. (1967) synthesized gyplure (VIII) and reported that it, as well as the compound proposed as the natural pheromone (III), had no biological activity (Section IV,C). Further interpretation of biological data resulting from use of these materials must await clarification of the validity of the proposed identifications and of the biological activity of synthesized chemicals.

   b. Cabbage looper.
   A small area (1/4-acre plots) field trial was designed to test the male inhibition principle (Gaston et al., 1967; Shorey et al., 1967). The trial was based on laboratory evidence that (1) males must respond to the sex pheromone emitted by a female to be able to mate with the female, (2) the BR_{20}, defined here as the lower threshold for male behavioral response to the female sex pheromone, is about 2×10^{-14} g/liter of air (60 molecules/mm^3), and (3) continual exposure to low concentrations of pheromone inhibits responsiveness of males to subsequent higher concentrations. An array of 100 pheromone evaporators was set up in the

279

plots to provide a concentration of about 10^{-10} g/liter of air (about 10^4 higher than behavioral threshold). The evaporators were reservoirs containing the pheromone and therefore had a relatively constant and reproducible evaporation rate (except for temperature dependence). Less than 0.2 g/acre/night of synthetic pheromone was expended. The success of this trial in preventing any wild males from orienting to caged females in the centers of the plots demonstrates the feasibility of the male inhibition technique.

    c. Discussion of male inhibition technique.

Once the essential biological parameters--including female release rates, times and seasons; moth flight ranges; the importance of aggregation sites for mating; and male responsiveness thresholds--are known, the male inhibition technique presents mostly an engineering problem (Wright, 1965).

It appears essential that evaporation rates from the substrates used should be determined for any chemicals intended for male inhibition experiments. For efficient control of a lepidopterous pest in the field, a device is needed that releases pheromones at a constant rate, compensating for the effect of varying temperature on the evaporation rate and having a mechanism to limit pheromone release to that time of day when mating occurs.

The production cost of many synthetic sex pheromones may be too high to allow expenditure of quantities near 0.2 g/acre/day. In the search for other, nonpheromone chemicals that inhibit responses of male moths to their natural sex pheromones, the chemical to be used is the one that exerts the greatest biological effect per unit cost.

VI. Summary

Several specific areas of recent progress in research on the sex pheromones of Lepidoptera have been examined.

A basic essential for many biological and chemical studies is a quantitative bioassay which is sensitive small amounts of pheromone and is reproducible.

A variety of hypotheses have been proposed in the literature to account for the aerial orientation of a male moth to a pheromone-releasing female. It is suggested that the orientation behavior of the male is based on micro-anemotaxis, fortified by the disrupted pheromone concentrations in the turbulent air masses through which he flies.

The great array of different pheromones used by the various lepidopterous species indicates that an extensive chemical evolution has occurred. It is proposed that females of each species produce a variety of related chemicals in their pheromone glands, with one of the chemicals selected out for mass production as the sex pheromone. One of the minor chemicals may be selected for by strong evolutionary pressures as the different sex pheromone in a newly evolved, reproductively isolated species.

Behavioral control of a lepidopterous species based on luring male moths to pheromone-emitting traps may not be feasible. The innate behavior of males may prevent them from approaching traps that release sufficient quantities of pheromone to out-compete wild pheromone-releasing females. Depending on cost factors, it may be possible to control some species by permeating air over a geographic area with enough pheromone to cause the additional increment contributed by wild females to be nonperceptible to males.

Acknowledgement

Original research reported here was supported in part by U.S. Public Health Service Grant No. GM 11524.

VII. References

Alexander, R. D. (1964). Symp. Roy. Entomol. Soc. London
     2, 78.
Ambros, W. (1938). Zentr. Ges. Forstwesen 64, 209.
Babson, A. L. (1963). Science 142, 447.
Berger, R. S. (1966). Ann. Entomol. Soc. Am. 59, 767.
Berger, R. S., and Canerday, T. D. (1968). J. Econ. Entomol.
     61, 452.
Berger, R. S., McGough, J. M., Martin, D. F., and Ball, L. R.
     (1964). Ann. Entomol. Soc. Am. 57, 606.
Beroza, M. (1967). J. Econ. Entomol. 60, 875.
Beroza, M., and Jacobson, M. (1963). World Rev. Pest
     Control 2, 36.
Block, B. C. (1960). J. Econ. Entomol. 53, 172.
Boeckh, J., Kaissling, K. E., and Schneider, D. (1965).
     Cold Spring Harbor Symp. Quant. Biol. 30, 263.
Bossert, W. H., and Wilson, E. O. (1963). J. Theoret. Biol.
     5, 443.
Burgess, E. D. (1964). Science 141, 526.
Butenandt, A., and Hecker, E. (1961). Angew. Chem. 73, 349.
Butenandt, A., Beckmann, R., Stamm, D., and Hecker, E.
     (1959). Z. Naturforsch. 146, 283.
Butenandt, A., Beckmann, R., and Hecker, E. (1961a). Z.
     Physiol. Chem. 324, 71.
Butenandt, A., Beckmann, R., and Stamm, D. (1961b). Z.
     Physiol. Chem. 324, 84.
Butt, B. A., and Hathaway, D. O. (1966). J. Econ. Entomol.
     59, 476.
Callahan, P. S. (1965). Ann. Entomol. Soc. Am. 58, 727.
Daykin, P. N., Kellogg, F. E., and Wright, R. H. (1965).
     Can. Entomologist 97, 239.
Dethier, V. G. (1957). Surv. Biol. Progr. 3, 149.
Doane, C. C. (1968). Ann. Entomol. Soc. Am. 61, 768.
Eiter, K., Truscheit, E., and Boness, M. (1967). Liebigs
     Ann. Chem. 709, 29.
Flaschenträger, B., and Amin, El. S. (1950). Nature 165,
     394.
Gaston, L. K., and Shorey, H. H. (1964). Ann. Entomol. Soc.
     Am. 57, 779.
Gaston, L. K., Shorey, H. H., and Saario, C. A. (1967).
     Nature 213, 1155.
Gaston, L. K., Shorey, H. H., and Saario, C. A. (1969). J.
     Econ. Entomol. (in press).

Götz, B. (1951). Experientia 7, 406.
Graham, H. M., Martin, D. F., Ouye, M. T., and Hardman, R. M.
    (1966). J. Econ. Entomol. 59, 950.
Heath, J. E., and Adams, P. A. (1967). J. Exp. Biol. 47,
    21.
Jacobson, M. (1964). Am. Heart J. 68, 577.
Jacobson, M. (1965). "Insect Sex Attractants." Wiley
    (Interscience), New York.
Jacobson, M. (1966). Ann. Rev. Entomol. 11, 403.
Jacobson, M., and Jones, W. A. (1962). J. Org. Chem. 27,
    2523.
Jacobson, M., Beroza, M., and Jones, W. A. (1961). J. Am.
    Chem. Soc. 83, 4819.
Jacobson, M., Toba, H. H., De Bolt, J., and Kishaba, A. N.
    (1968). J. Econ. Entomol. 61, 84.
Jones, W. A., Jacobson, M., and Martin, D. F. (1966).
    Science 152, 1516.
Karlson, P., and Butenandt, A. (1959). Ann. Rev. Entomol.
    4, 39.
Kellogg, F. E., Frizel, D. E., and Wright, R. H. (1962).
    Can. Entomologist 94, 884.
Kennedy, J. S. (1939). Proc. Zool. Soc. (A) 109, 221.
Kettlewell, H. B. D. (1946). Entomologist 79, 8.
Knipling, E. F., and McGuire, J. U., Jr. (1966). U.S.
    Dept. Agr. Inform. Bull. 308, 1.
Laithewaite, E. R. (1960). Entomologist 93, 113, 133, 232.
Mayr, E. (1963). "Animal Species and Evolution." Harvard
    Univ. Press, Cambridge.
Moore, B. P. (1967). Science J. 3(9), 40.
Neuhaus, W. (1965). Z. Vergleich. Physiol. 49, 475.
Read, J. S., Warren, F. L., and Hewitt, P. H. (1968).
    Chem. Comm. 1968, 792.
Roelofs, W. L., and Arn, H. (1968). Nature 219, 513.
Schneider, D. (1962). J. Insect Physiol. 8, 15.
Schneider, D. (1965). Symp. Soc. Exp. Biol. 20, 273.
Schwinck, I. (1953). Z. Vergleich. Physiol. 35, 167.
Schwinck, I. (1955). Z. Vergleich. Physiol. 37, 439.
Schwinck, I. (1958). Proc. 10th Intern. Congr. Entomol.,
    Montreal, (1956) 2, 577.
Sekul, A. A., and Sparks, A. N. (1967). J. Econ. Entomol.
    60, 1270.
Shorey, H. H. (1964). Ann. Entomol. Soc. Am. 57, 371.
Shorey, H. H. (1966). Ann. Entomol. Soc. Am. 59, 502.
Shorey, H. H., and Gaston, L. K. (1964). Ann. Entomol. Soc.
    Am. 57, 775.

Shorey, H. H., and Gaston, L. K. (1965a). Ann. Entomol. Soc. Am. 58, 604.

Shorey, H. H., and Gaston, L. K. (1965b). Ann. Entomol. Soc. Am. 58, 833.

Shorey, H. H., and Gaston, L. K. (1967). In "Pest Control: Biological, Physical and Selected Chemical Methods." (Eds., Kilgore, W. W., and Doutt, R. L.) Academic Press, N.Y.

Shorey, H. H., and Gaston, L. K. (1969). Ann. Entomol. Soc. Am. (in press).

Shorey, H. H., Gaston, L. K., and Roberts, J. S. (1965). Ann. Entomol. Soc. Am. 58, 600.

Shorey, H. H., Gaston, L. K., and Saario, C. A. (1967). J. Econ. Entomol. 60, 1541.

Shorey, H. H., Gaston, L. K., and Jefferson, R. N. (1968). In "Advances in Pest Control Research." (Ed., Metcalf, R. L.) Interscience Publishers, N. Y. 8, 57.

Steinbrecht, R. A. (1964). Z. Verleich. Physiol. 48, 341.

Teale, E. W. (1961). "The Insect World of J. Henri Fabre." Dodd, Mead, N. Y.

Traynier, R. M. M. (1968). Can. Entomologist 100, 5.

Waters, R. M., and Jacobson, M. (1965). J. Econ. Entomol. 58, 370.

Wilson, E. O. (1963). Sci. Am. 208, 100.

Wolf, W. W., Kishaba, A. N., Howland, A. F., and Henneberry, T. J. (1967). J. Econ. Entomol. 60, 1182.

Wright, R. H. (1958). Can. Entomologist 90, 81.

Wright, R. H. (1964a). "The Sense of Smell." George Allen and Unwin, London.

Wright, R. H. (1964b). Nature 204, 121.

Wright, R. H. (1965). Bull. At. Scientists 21, 1.

METHODOLOGY FOR ISOLATION AND
IDENTIFICATION OF INSECT PHEROMONES-
EXAMPLES FROM COLEOPTERA

Robert M. Silverstein

Department of Chemistry
State University College of Forestry
at Syracuse University
Syracuse, New York

Table of Contents

ROBERT M. SILVERSTEIN

I. Introduction

The organic chemist finds an irresistible appeal in the chemistry of insect pheromones. The problems of isolation and identification of the active compounds are challenging, but yield to the modern techniques at his disposal. The phenomena he deals with are biologically significant and intellectually satisfying to the point that he is tempted to make simplistic interpretations of his observations. The best safeguard lies in close collaboration with a sophisticated biologist, who knows that:

> To him who in the love of Nature holds
> Communion with her visible forms, she speaks
> A various language;---.
> Thanatopsis

Finally the results have promise in the control of insects, and without further insensate despoliation of our ecology.

To illustrate the methodology for isolation and identification of insect pheromones, I shall describe our studies on the sex attractants of two bark beetles, which are forest insect pests, and the black carpet beetle, a pest in dwellings and in grain storage and processing facilities. Unlike the sex attractants that have been reported thus far for Lepidoptera, which are all long chain unsaturated fatty alcohols or esters, the compounds isolated from these three Coleoptera vary widely in chemical structure. The bark beetle studies (Ips confusus and Dendroctonus brevicomis) were done in collaboration with Dr. D. L. Wood, Department of Entomology and Parasitology, University of California, Berkeley. The carpet beetle study (Attagenus megatoma) was carried out with Dr. W. E. Burkholder, Department of Entomology, University of Wisconsin, Madison.

A rigorous approach to identification of insect pheromones involves the following steps:

1. Development of a bioassay suitable for monitoring the isolation steps.

2. Production of large amounts of starting material - i.e., mass rearing.

3. A series of progressively refined isolation

steps, each monitored by bioassay, until individual active compounds are obtained in a high state of purity.

4. Structure determination of the individual active compounds.

5. Confirmation of postulated structures by comparison of their properties with those of authentic samples.

6. Confirmation of biological activity of synthesized compounds in the laboratory and in the field.

7. Field applications.

Since field tests require large quantities of material, a convenient laboratory bioassay is a sine qua non for monitoring isolation steps. This bioassay must reflect the behavior of insects in the field. Obviously there is some risk in using, for example, a walking response in the laboratory as a measure of a flying response in the field. This calculated risk was assumed in our work with bark beetles, and satisfactory correlations were made (see chapter by Wood).

The past two decades have seen remarkable developments in chromatographic techniques for isolation of compounds from complex mixtures. Yet even with these techniques, the problems of isolating minute amounts of biologically active products from plants or animals (or parts thereof) or from such conversion products as frass are still formidable. Active sites in chromatographic substrates can cause chemical changes during column and thin-layer chromatography. Gas chromatography involves the additional hazard of thermal degradation. But these hazards are slight indeed compared with those of the classical chemical manipulations.

An even more serious problem than instability is the extraordinary activity of many insect pheromones. Many an investigator has achieved what he considered to be gas chromatographic homogeneity (i.e., a single, apparently symmetrical peak), and has identified the compound responsible for the peak by comparison with an authentic synthetic sample, only to find that the synthetic sample is

inactive. The active compound, in fact, is a minor component buried under the large peak. The worst possible situation exists if the active minor component is a closely
related isomer of the inactive major component, which is
quite likely if a series of manipulations fails to separate one from the other. Then it is possible that a synthesis sequence may produce both isomers, and that again there
is no separation. The chemist, then, has played the game
according to the rules, and has arrived at the wrong answer.
His only recourse is exhaustive purification of both the
isolated and synthesized compounds monitored by a quantitative (at least semi-quantitative) bioassay at each stage.
The situation is further complicated by masking and synergistic effects.

Developments in techniques for isolating pure compounds
from mixtures have been paralleled by an even more remarkable development in spectrometric instrumentation for determination of chemical structure. Organic compounds can frequently be identified from the complementary information
afforded by four spectra: mass, infrared, nuclear magnetic,
and ultraviolet (Silverstein and Bassler, 1967). Identification at the level of several milligrams is routine;
identification at the microgram level is possible, but
physical manipulations become a major preoccupation. The
limiting factor is usually the minimum quantity needed for
a nuclear magnetic resonance spectrum, about 50 to 100
micrograms. Not all molecules yield so easily, and chemical
manipulations may be necessary. But the spectral information permits intelligent selection of chemical treatment,
and spectral information can be obtained on the resulting
products.

The "moment of truth" for the organic chemist comes
when he matches the spectra of his isolated compound against
those of an authentic sample. The entomologist who has
carried out the original behavioral studies, devised the
bioassay, and mass-reared the insects makes the final con-
tribution when he learns to use the synthetic compounds as
tools for survey and control purposes.

II. Ips confusus (LeC.)

Bark beetles have been recorded killing the equivalent
of five billion board feet of timber or as much as six times
that caused by fire in one year. There are no effective
direct control methods.

The attack on a tree by many kinds of bark beetles consists of two phases: an initial attack by a few beetles, followed by a massive secondary invasion that kills the tree. As the initial attackers bore into the tree to construct a nuptial chamber, they expel frass--a mixture of fecal pellets and wood fragments. This frass (actually the fecal pellets) contains an attractant that triggers the secondary invasion. In the case of I. confusus, the attractant is produced by the male, and it attracts both females and males. Three females mate with each male.

Frass was produced in Dr. Wood's laboratory by newly-emerged male beetles that had been introduced into cut ponderosa pine logs (Wood et al., 1966). The frass was extracted in a Waring blender with benzene. The benzene was removed by distillation, and the extracted material was subjected to short-path, high-vacuum distillation. This was a powerful clean-up step; about 93% of the extractives did not distill and were discarded. The next step was silica gel chromatography, followed by gas chromatography on an SE-30 column of the active fraction. The active fraction from the SE-30 column was rechromatographed on a Carbowax 20 M column from which 5 fractions were collected-- none of which were active. However, when two of the fractions were recombined, most of the original activity was restored.

Eventually three terpene alcohols were isolated from 4.5 kg of frass, which represented the output of 20,000 male beetles (Silverstein et al., 1966a,b). None of the compounds were active by themselves, but a combination of two was necessary to elicit response in the laboratory bioassay. In the field, a mixture of all three compounds was the most attractive to flying insects (Wood et al., 1967, 1968).

Some indication of the tediousness of the isolation procedure may be gained by noting that six sequential fractionations by gas chromatography were needed to obtain a pure sample of one of the terpene alcohols. The postulated structures were confirmed by congruence of spectra with those of synthesized compounds, which also were biologically active.

Figure 1 presents the spectra for one of the compounds. All spectra displayed are those actually obtained on the isolated products. Let us go through the evidence for its identification in some detail. The mol. wt. is 154 (M peak in mass spectrum). The M+1 and M+2 peaks are too weak to

289

be accurately measured. The IR spectrum shows an O-H stretching band at 2.96 μ, an olefinic C-H stretching band at 3.23 μ, a conjugated C=C stretching band at 6.25 μ, and characteristic "vinyl" absorption at 10.1 μ and 11.2 μ. The presence of a geminal dimethyl group is suggested by the double peak at 7.21 μ and 7.29 μ. The UV spectrum suggests a conjugated double bond system (λ_{max}^{hexane} 226 mμ, ε 20,000). The vinyl proton peaks in the NMR spectra show a 1:4 ratio, and the downfield proton (τ 3.6) is split only by two geminal protons (J_{trans} 18 c/s, J_{cis} 12 c/s). The following moiety can be written:

$$CH_2$$
$$\parallel$$
$$-C-CH=CH_2$$

The molecular formula must be $C_{10}H_{18}O$. Given the context, we think in terms of a terpene alcohol.

Shaking the deuterochloroform solution with deuterium oxide caused the peak at τ 8.40 in the NMR spectrum to disappear. The deshielded multiplet at τ 6.2 (1 proton) can be tentatively ascribed to a proton geminal to the hydroxyl group. This assignment was confirmed by the shift of this proton to τ 4.9 on acetylation with acetic anhydride in pyridine at room temperature. The mass, IR, UV and NMR spectra of the acetate were in accord with acetylation without rearrangement of the rest of the molecule. The parent peak in the mass spectrum of the acetate was 196, and the M+1 and M+2 peaks were 13.4% and 1.2% for $C_{12}H_{20}O_2$ which provides confirmation for the assigned molecular formula of the alcohol, $C_{10}H_{18}O$.

The presence of an isopropyl group can be surmised from the strong peak in the mass spectrum at m/e 43, and from the double peak in the IR spectrum at 7.21 μ and 7.29 μ. The upfield doublets in the NMR spectrum peaks (6 protons) must represent the geminal methyl groups which have slightly different shift positions (τ 9.08 and τ 9.10); each methyl group is split by the vicinal CH, giving rise to two overlapping doublets (J 7 c/s). The five protons (excluding the OH proton) between ∿ τ 7.4 and ∿ τ 8.9 occur as two multiplets of two protons each, separated by a multiplet of one proton that must represent the methine proton of the isopropyl group. We thus have the following fragments which account for the molecular formula:

$$
\overset{\overset{\displaystyle CH_2}{\|}}{-C-CH=CH_2} \quad -\overset{|}{C}HOH \quad CH_3\overset{|}{C}HCH_3 \quad \overset{|}{C}H_2 \quad \overset{|}{C}H_2
$$

The vinylidene out-of-plane bending absorption could be accommodated under the broad IR band at 11.2 µ.

Decoupling the CHOH proton (τ 6.2) collapsed the downfield CH_2 pattern to a pair of doublets and also affected the upfield CH_2 pattern (Fig. 1); the CHOH group must therefore lie between the CH_2 groups. Attempts to decouple the very broadly split CH_3CHCH_3 proton were only partly successful; however, the methyl pattern was changed. Thus, the general character of the methyl groups was confirmed, and the upfield CH_2 group was located adjacent to the isopropyl group.

The optical rotation ($[\alpha]_D^{25°}$,c, 1 EtOH) was -17.5° ± 0.7°. The structure, therefore, is (-)-2-methyl-6-methylene -7-octen-4-ol.

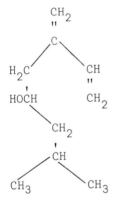

This structure satisfies all the spectrometric requirements. Deshielding of one of the methylene groups by a neighboring double bond and a hydroxyl group accounts for its downfield position in the NMR spectrum. The protons of this deshielded methylene group are not equivalent because of the adjacent asymmetric center; one proton is at τ 7.53, the other at τ 7.82. Each is split by the other, J_{gem} 14 c/s, and by the neighboring proton, J_{vic} 9 c/s and J_{vic} 4 c/s. The upfield methylene multiplet also comprises two nonequivalent methylene protons but the couplings are too complex for first-order interpretation. The non-equivalence of the methyl groups can be rationalized even though the asymmetric center is separated from the isopropyl moiety by a methylene group.

The principal cleavages in the mass fragmentation pattern might be expected on either side of the CHOH group. One of these cleavages (m/e 87) does account for a major peak. The base peak is m/e 68, which presumably arises from allylic cleavage with rearrangement of a hydrogen atom. The moderately intense peaks at m/e 136 and m/e 121 represent consecutive elimination of water and a methyl group. This sequence, commonly found in mass spectra of terpene alcohols, confirms m/e 154 as the parent peak.

The other two components of the I. confusus attractant were cis-verbenol and a compound similar to the one just described, but with a double bond in the 2-position.

III. Dendroctonus brevicomis (LeC.)

The western pine beetle (D. brevicomis) is one of the most destructive bark beetle pests in North America. In this insect, the monogamous female produces the sex attractant.

From the frass a single compound was isolated that was active by itself (Silverstein et al., 1968). A hydrocarbon fraction behaved synergistically to enhance its effect. The isolation procedure was similar to that described for I. confusus.

The spectra for the active compound, which we named brevicomin, are shown in Fig. 2. Identification was made on 2 mg isolated from 1.6 kg of frass.

The molecular formula from high resolution mass spectrometry is $C_9H_{16}O_2$. The molecule therefore contains either a triple bond, 2 double bonds, a double bond and a ring, or 2 rings. The infrared spectrum indicates that the oxygen atoms represent ether groups (absence of hydroxyl and carbonyl). The UV shows no conjugated unsaturation.

The NMR shows the methyl protons of an ethyl group (triplet at τ 9.13), a methyl group on a deshielded quaternary carbon (singlet at τ 8.70), and two individual protons at τ 6.02 and τ 6.22 which must be on carbon atoms adjacent to oxygens.

In the absence of evidence for a triple bond or double bonds, bicyclic structures should be considered.

Epoxide and peroxide structures were ruled out because the compound did not react with lithium aluminum hydride. We carried out catalytic hydrogenalysis at 250°C on a 50 μg sample in a Beroza carbon-skeleton determinator (Beroza and Sarmiento, 1963; Brownlee and Silverstein, 1968), and identified the product by mass spectrometry as nonane. Therefore the ethyl and methyl groups must be at the ends of the opened molecule.

$$\underline{CH_3}\text{-C-C-C-C-C-C-}\underline{CH_2CH_3}$$

Since the methyl group is attached to a quaternary carbon, and is deshielded, at least one of the oxygen atoms must be on the carbon adjacent to the methyl group.

$$\underline{CH_3}\text{-C}\overset{\diagup O}{}\text{C-C-C-C-C-}\underline{CH_2CH_3}$$

There is only one proton on each of two carbon atoms adjacent to the oxygen atoms. This really limits the number of possible structures. We were forced to write a bicyclic ketal.

$$\underline{CH_3}\text{-C}\overset{\diagup O}{\underset{\diagdown O}{}}\text{C-C-C-C-C-}CH_2CH_3$$

If the ethyl group is a side chain on a ring, one oxygen atom must be joined to the number 7 carbon, and in order to account for the downfield triplet we have to replace at least one of the protons on the number 6 carbon by connecting with the other oxygen atom.

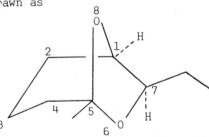

$$^1CH_3{}^2C\text{---}^3C\text{-}^4C\text{-}^5C\text{-}^6C\text{-}^7C\text{-}C^8H_2{}^9CH_3$$

This can be redrawn as

exo-7-ethyl-5-methyl-6,8-dioxabicyclo [3.2.1] octane

293

The proton on the number 6 carbon (now 1) is at right angles to the number 7 proton, and therefore the coupling constant is close to zero.

The spectral properties of this compound were congruent with those of a synthesized compound which showed biological activity in the laboratory and the field.

It is interesting to note that the endo- structure is also present in frass, but is inactive when presented alone to males in the laboratory olfactometer.

Myrcene, isolated and identified from the hydrocarbon fraction, enhanced the activity of exo-brevicomin for males.

IV. Attagenus megatoma (F.)

The black carpet beetle (A. megatoma) is a pest in homes where the larvae eat clothing, rugs, fur and feathers. A large part of the damage attributed to moths is really caused by the black carpet beetle. It is also a serious pest in stored grain and flour. The sex attractant elaborated by the female both attracts and excites the male (Burkholder and Dicke, 1966).

The female beetles were extracted with benzene in a Waring blender (Silverstein et al., 1967). The benzene was removed by distillation, and the residue was distilled under high vacuum in a short-path distillation apparatus. The distillate was fractionated on a silica gel column. The active fraction therefrom did not survive gas chromatography. We then established that the active fraction of the distillate could be extracted from an ether solution with an ice-cold 0.1 N sodium hydroxide solution, which on reacidification and extraction with ether gave a mixture of acidic compounds containing the active compound. This mixture was chromatographed on silica gel, and the active fraction therefrom was chromatographed on an anion-exchange column. Treatment of this active fraction with diazomethane gave a mixture of methyl esters that were separated by gas chromatography on an SE-30 column. Another separation on SE-30 of the collected active fraction followed by further fractionation on Carbowax 20 M gave a pure compound, which was identified from the spectra shown in Fig. 3. Of course, the gas chromatographic fractionation of the methyl ester was monitored by saponifying an aliquot of the fractions. Each unmated female beetle contained about 0.5 µg of the attractant.

The identification was accomplished on about 4 mg of the methyl ester (Fig. 3). From the high resolution mass spectrum, the molecular formula is $C_{13}H_{23}COOCH_3$ which allows a triple bond, 2 double bonds, 2 rings, or a ring and a double bond. The ultraviolet spectrum indicates two conjugated double bonds that are not conjugated with the carbonyl group. We have only to locate the double bonds and to determine whether we have a straight or a branched chain.

The NMR shows four olefinic protons between $\sim \tau$ 3.5 and $\sim \tau$ 4.8. The carbomethoxy protons are seen as a singlet at τ 6.4. The doublet at τ 7.0 represents two protons that must be deshielded both by a carbonyl group and a double bond. We can therefore write

$$-CH=CH-CH=CH-\underline{CH_2}-COOCH_3$$

The quartet at τ 7.8 represents a methylene group deshielded by a double bond. Hence, a partial structure is

$$-CH_2-\underline{CH_2}-CH=CH-CH=CH-CH_2-COOCH_3$$

There are 12 protons (6 methylene groups) at τ 8.7 and three protons under the characteristically distorted triplet at τ 9.1. We therefore write a straight chain

$$CH_3(CH_2)_7CH=CH-CH=CHCH_2COOCH_3$$

In the infrared spectrum, we see the characteristic cis-trans absorption at 10.20 μ and 10.55 μ, but we cannot tell which of the two possible isomers we have. The four possible isomers were synthesized. The spectra of the isolated compound matched those of the trans-3,cis-5 carboxylic acid, which we have named megatomoic acid. The synthetic material is active in the laboratory bioassay and in field tests.

Acknowledgments

The isolation, identification, and synthesis of the sex attractants were done at the Stanford Research Institute by J. O. Rodin, R. G. Brownlee, G. C. Reece, and T. E. Bellas. The successful outcome of these investigations is a tribute to the skill and enthusiasm of this group.
The bark beetle studies were supported by the US Forest

Service, and the black carpet beetle work by the Market
Quality Research Division, US Department of Agriculture.

V. References

Beroza, M., and Sarmiento, R. (1963). Anal. Chem. 35,
1353.
Brownlee, R. G., and Silverstein, R. M. (1968). Anal. Chem.
40, 2077.
Burkholder, W. E., and Dicke, R. J. (1966). J. Econ.
Entomol. 59, 540.
Silverstein, R. M., Rodin, J. O., and Wood, D. L. (1966a).
Science 154, 509.
Silverstein, R. M., Rodin, J. O., Wood, D. L., and Browne,
L. E. (1966b). Tetrahedron 22, 1929.
Silverstein, R. M., and Bassler, G. C. (1967). "Spectro-
metric Identification of Organic Compounds". John
Wiley & Sons, Inc., N. Y., 2nd edition.
Silverstein, R. M., Rodin, J. O., Burkholder, W. E., and
Gorman, J. E. (1967). Science 157, 85.
Silverstein, R. M., Brownlee, R. G., Bellas, T. E., Wood,
D. L., and Browne, L. E. (1968). Science 159, 889.
Wood, D. L., Browne, L. E., Silverstein, R. M., and Rodin,
J. O. (1966). J. Insect Physiol. 12, 523.
Wood, D. L., Stark, R. W., Silverstein, R. M., and Rodin,
J. O. (1967). Nature 215, 206.
Wood, D. L., Browne, L. E., Bedard, W. D., Tilden, P. E.,
Silverstein, R. M., and Rodin, J. O. (1968). Science
159, 1373.

CONTROL OF INSECT BEHAVIOR

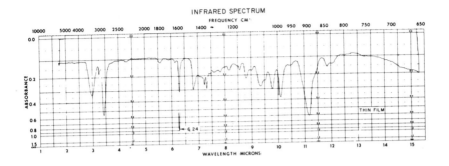

INFRARED SPECTRUM

MASS SPECTRAL DATA (RELATIVE INTENSITIES)

m/e	% OF BASE PEAK	m/e	% OF BASE PEAK	m/e	% OF BASE PEAK
38	4	53	25	77	3
39	37	55	6	79	6
40	12	57	17	80	4
41	70	58	7	85	15
42	10	65	4	87	10
43	66	66	5	91	3
44	18	67	35	93	8
45	26	68	100	97	2
50	3	69	64	98	2
51	6	70	5	121	2
52	4	71	8	136	2
				154 (M)	1.1

UV DATA

λ_{max} 226 mμ
ϵ_{max} 20,000
SOLVENT HEXANE

OPTICAL ROTATION

$[\alpha]_D^{25°}$ −17.5°±0.7°

c = 1 IN ETHANOL

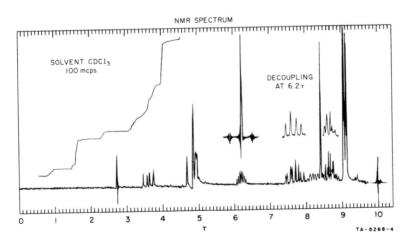

NMR SPECTRUM

SOLVENT CDCl₃
100 mcps

DECOUPLING
AT 6.2 τ

TA-5268-4

Figure 1.--Spectral data for the Ips confusus attractant.

297

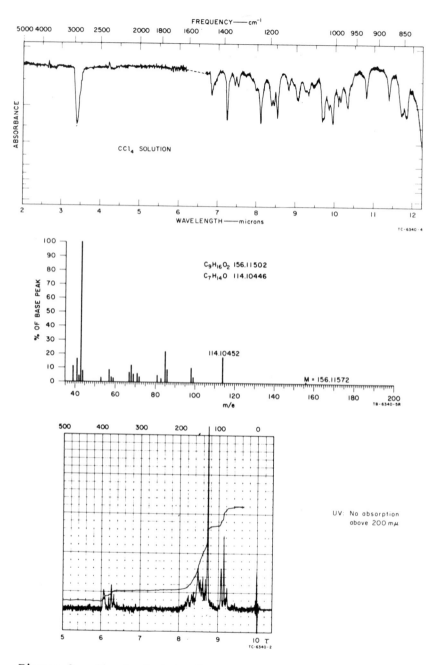

Figure 2.--Spectral data for the Dendroctonus brevicomis attractant.

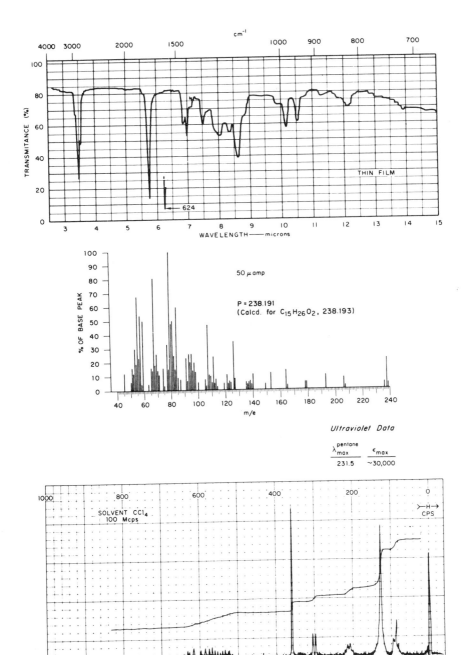

Figure 3.--Spectral data for the Attagenus megatoma attractant.

PHEROMONES OF BARK BEETLES

David L. Wood

Department of Entomology and Parasitology
University of California, Berkeley

Table of Contents

I. Introduction

Person's (1931) early work with the western pine beetle, Dendroctonus brevicomis LeConte, established the foundation for current research on bark beetle pheromones. He proposed a host selection theory which states that the initial attraction of beetles to susceptible ponderosa pines (Pinus ponderosa Laws.) is caused by "...the formation and escape of the volatile aldehydes or esters which are a by-product of a respiratory fermentation resulting from abnormal enzyme activity in subnormal trees... After a few attacks are made, a second stronger attraction is started by yeasts introduced by the attacking beetles. This secondary attraction is probably strong enough to attract beetles for a considerable distance with the result that the tree is usually heavily attacked and killed." In the pioneering study by Anderson (1948) the secondary attraction was attributed to the males of Ips pini (Say) boring in P. banksiana Lamb. This was later verified in a study of I. confusus (LeConte) attack behavior (Wood and Vité, 1961). Subsequently, the secondary attraction phenomenon has been described for many scolytid species in a number of genera, i.e., Carphoborus (Chararas, 1966), Dendroctonus (Dyer and Taylor, 1968; McCambridge, 1967; McMullen and Atkins, 1962; Pitman et al., 1968; Rudinsky, 1963; Tsao and Yu, 1967; Vité and Gara, 1962; Vité et al., 1964), Ips (Bakke, 1967; Vité and Gara, 1962; Vité et al., 1964; Wilkinson, 1964), Pityogenes (Vité, 1965), Pityophthorus (Vité, 1965), Scolytus (Goeden and Norris, 1964; Meyer and Norris, 1967), and Trypodenron (Chapman, 1966; Rudinsky and Daterman, 1964). The literature accumulated during the past 8 years testifies to the interest of scolytid biologists in this phenomenon. Such interest is undoubtedly motivated, at least in part, by the potentialities for behavioral control of this destructive group of forest insects. This paper will emphasize studies of I. confusus and D. brevicomis performed at the University of California in cooperation with the Stanford Research Institute and the U. S. Forest Service, and is not intended to be a complete review of research on bark beetle pheromones.

II. Review

In the Scolytidae, beetles of only one sex select the individual host tree in which mating and subsequent oviposition will occur. In the monogamous genus Dendroctonus, the entrance tunnel is excavated by the female, while in the polygamous genus Ips, the male performs this function.

After arriving on the tree, the I. confusus male selects a gallery site in a bark crevice and bores the entrance tunnel through the outer bark into the phloem tissue where he then excavates a "nuptial" chamber. Usually 3 females join the male, each constructing an egg gallery in the phloem and outer xylem.

The host colonization pattern of I. confusus in P. ponderosa (Wood and Vité, 1961) consists of 3 phases:
1. initial attack - the occurrence of a few male attacks on a living tree or one recently cut;
2. mass attack - an abrupt increase in the number of male attacks;
3. terminal attack - a gradual decline in the number of new male attacks.
Colonization can occur over a period of a few days to several weeks depending upon population levels and weather conditions.

The successful development of a laboratory bioassay (Wood, 1962; Wood and Bushing, 1963) of the olfactory response to the male-produced attractant permitted a precise localization of its source. A multiple choice olfactometer, consisting of 5 sample positions, was adapted from an arena design. Air was forced from a squirrel-cage fan through Tygon plastic tubing into the rear of plastic receptacles, over the samples, and then exhausted through a screened exit port. Thus, volatile materials could be directed at floor level to a common exhaust point where groups of beetles were released. Their movement to the screened exit port containing the attractant material and their klinotactic behavior in the olfactometer established the basis of the bioassay.

The secondary attractant (variously termed "population-aggregating" or "sex" pheromone; see exchange between Vité (1967) and Wood et al. (1967b)) was first localized in the frass material (primarily a mixture of phloem fragments and excrement pellets) produced by the male beetle boring in freshly cut ponderosa pine (Wood, 1962). It occurs in the frass 9-12 hr following introduction of the males into preformed entrance tunnels (Wood and Bushing, 1963). The pheromone is present on (Pitman et al., 1965; Vité et al., 1963) and throughout (Wood et al., 1966) the dense fecal pellet. Although the pheromone is concentrated in the hindgut (Borden, 1967; Pitman et al., 1965; Vité et al.,

1963), the exact site of elaboration remains unknown. The
hindgut epithelium has been implicated (Pitman and Vité,
1963) but subsequently denied (Pitman et al., 1965).
Pheromone production occurs only after feeding commences,
indicating that either a precursor is ingested and metabolized to the attractant or that metabolism of food material
causes secretory activity in specialized cells (Wood et al.,
1966). Glucose-supplemented diets appeared to favor pheromone production (Pitman, 1966).

Various aspects of the behavior of I. confusus relating
to the production of and response to male attractant have
been studied in the laboratory (Borden, 1967). Female response to male frass reached a low of 24% in January and
February and rose to a peak of 76% in May and June. Response to the same benzene extract of male frass decreased
moderately during autumn and winter, indicating a lowered
response ability for overwintering populations. After brief
exposure periods (about 30 sec), even non-responders oriented
to male attractant at significantly higher levels than the
total original population. Confinement in an atmosphere
containing male frass odor for more than 15 min caused a
highly significant reduction in female response to male
frass extract. Both contact with male frass and with
attractant-laden air reduced take-off attempts of both
sexes to near zero. Reproducing females showed a greatly
reduced response to male attractant. Single males produced
frass throughout the day but maximum production was coinci-
dent with the time of peak female emergence and flight
(Cameron and Borden, 1967).

In both sexes of I. confusus, removal of the antennal
clubs eliminated a positive response to the pheromone
(Borden and Wood, 1966). The response of the females after
various portions of the club were removed or covered indicates that the sensilla trichodea are receptors of the
pheromone.

The pheromones of I. confusus were isolated from 4.5 kg
of frass produced by male beetles (Wood et al., 1966). Infested logging slash from ponderosa pine was caged in the
greenhouse; after emergence, the males were separated from
the females and introduced into preformed entrance tunnels
on freshly cut ponderosa pine logs. The outer bark was
shaved smooth to permit frass to fall unimpeded to the cage
floor. The frass was swept from the cage floor into bottles
and the air displaced from unfilled space with nitrogen.

The bottles were then stored under refrigeration. Frass
was produced from each log for 15-18 days following introduction of the males. The average daily frass output per
male for the first 15 days was 9.4 mg. A detectable response
was evoked by dilutions of benzene extracts equivalent to
1×10^{-8} g of frass, which is about that produced by one
male in 1/3 sec.

Three terpene alcohols were identified and synthesized
(Silverstein et al., 1966a,b) and are believed to be the
principal components of the secondary attractant that evokes
the mass attack. Compound I is (-)-2-methyl-6-methylene-7-
octen-4-ol; compound II, (+)-cis-verbenol; and compound III,
(+)-2-methyl-6-methylene-2,7-octadien-4-ol. In the laboratory bioassay, a typical olfactory response was elicited by
each of 3 mixtures: 1 µg of I with 0.01 µg of II, 1 µg of
I with 1 µg of III (Silverstein et al., 1966a), and 0.2 µg
of I with 0.01 µg of II and 0.01 µg of III (Wood et al.,
1967a). The compounds were inactive singly at these levels:
I at 100 µg, II at 20 µg, and III at 100 µg.

Attempts to isolate and identify I. confusus attractants
from hindguts dissected from feeding males have apparently
been unsuccessful (Renwick et al., 1966; Pitman et al.,
1966).

Utilizing the above methodology, the principal pheromone
occurring in the frass produced by female D. brevicomis boring in ponderosa pine has been recently isolated, identified,
and synthesized (Silverstein et al., 1968). This compound,
called exo-brevicomin (exo-7-ethyl-5-methyl-6,8-dioxabicy-
clo[3.2.1]octane), unlike the I. confusus pheromones, is
active alone. However, the activity of this compound is
enhanced by the terpene hydrocarbon myrcene, and other
components not yet identified.

Trans-verbenol is a major volatile component in the
aqueous extract of hindguts of newly emerged female D. brevicomis and D. frontalis Zimmerman; verbenone occurs in the
hindgut of the males (Renwick, 1967). Although trans-verbenol is claimed to be an active component of the D. frontalis pheromone (Vité and Pitman, 1968), the role of this
compound and of verbenone in the olfactory response of the
above species remains to be established. Trans-verbenol
has also been isolated and identified from an aqueous extract of hindguts dissected from female D. ponderosae
Hopkins boring in western white pine, P. monticolae Dougl.
(Pitman et al., 1968). When trans-verbenol dissolved in

petroleum ether was sprayed on female-infested billets,
there was about a 12-fold increase in catch compared to female-infested billets sprayed with petroleum ether only.

III. Field Experiments

The synthetic I. confusus pheromones were field-tested
during 1966 and 1967 (Wood et al., 1968). Experiments were
performed in a gently rolling, treeless brushfield to avoid
unknown, competing sources of attraction and unpredictable
air movements. Insect traps were placed on 20 m centers
in 2 lines 400 m apart and perpendicular to the prevailing
winds. Each treatment was replicated once in each line,
and the trap positions were assigned randomly at the beginning of each experiment.

The traps consisted of hardware-cloth (0.64 to 0.96 cm
mesh) cylinders, 30 cm high and 12 cm dia, each fastened
to the end of a 1.5 m rod driven into the ground. The
cylinders were dipped in warm "Stickem Special", and
most of the squares remained open. Air was metered from
a portable pressurized tank through a charcoal filter,
then through nylon tubing (0.32 cm O.D.) to an aluminum
tube containing the pheromones. The flow rate at the exit
port was 50 cm^3/min. The exit port was positioned in the
center of the trap. One bolt infested with 20 males was
placed in each line as the standard source of natural
attractant.

The synthetic pheromones were confined to an aeration
tube consiting of 1 cm O.D. aluminum tubing. A solution
of 0.5 g of Carbowax 20M in methanol was added to 9.5 g
of Chromosorb A, and most of the methanol was evaporated.
A solution of 1.5 mg of compound I in pentane was added
to 7 ml of the above substrate. The excess pentane was
evaporated, and the mixture was poured into a 15 cm length
of tubing (section A). A second 15 cm tube (section B)
was filled with a solution of 1.5 mg of compound I, 1 mg of
compound II, and 1 mg of compound III in pentane applied to
the substrate in the same manner. Glass wool plugs were
inserted into the ends of the tubes, and they were joined
together with a short piece of rubber tubing. Extract
equivalent to 2 g of male frass was applied to the substrate
in the same manner and confined to a 15 cm tube. Air was
passed through the tube from section A to section B. Because compound I traveled through this substrate twice as

306

fast as compounds II and III, it was distributed over twice the length.

In the first successful experiment, 57 beetles were attracted to the ternary synthetic mixture and 72 to the male-infested bolt in the first 6 hr of the test (Table I). Similar results were recorded when the experiment was repeated on the following day. The individual components and all possible combinations were then exposed under the same conditions. I. confusus was again highly responsive to the ternary mixture on 4 different days (Table II). This mixture appeared to be competitive with male-infested bolts. A mixture of compounds I and III attracted only 3 beetles. No other compound or combination of compounds was active. Extracts of male frass remained attractive for a longer period than did the synthetic compounds (Table I). In earlier trials, solutions of compounds I, II, and III in hexane or heptane, evaporated from cotton or paper wicks, elicited a very weak response. It is not understood how the solvent interfered with the attractant response.

Table I

Numbers of Ips confusus Trapped in Response to Extract of Male Frass, Male-infested Bolts, and Synthetic Pheromones (Madera County, California, September 1967)

Treatment	Numbers trapped[1]			
	Sept. 13		Sept. 14	
	A	B	A	B
Compounds I, II, and III	57	58	40	41
Extract of 2 g male frass	13	28	15	21
Male-infested bolts	72	108	59	83
Blank	0	0	0	0

[1] Numbers trapped at each time period represents the sum of 2 replications. A, 6 hr after start; B, 12 hr after start.

DAVID L. WOOD

Table II

Numbers of Ips confusus Trapped in Response to Synthetic
Pheromones Presented Individually and in Combination
(Madera County, California, September 1967)

Treatment	Numbers trapped[1]			
	Sept. 20	Sept. 26	Sept. 21	Sept. 25
I, II, and III	80	52	102	16
I	0	0		
II	0	0		
III	0	0		
I and II			0	0
I and III			3	0
II and III			0	0
Infested bolt	70	80	151	27
Blank	0	0	0	0

[1] Numbers trapped each day represents the sum of 2
replications.

The response of Enoclerus lecontei (Wolcott) (Coleoptera: Cleridae) to the synthetic pheromones was a unique
aspect of this study. Only compound II failed to elicit
a response from this predator (Table III). Although few
predators were caught, the ternary mixture appeared to be
as attractive as the male-infested bolt. The arrival of
adult predators on the host tree has been observed to coin-
cide with the bark beetle mass attack (Berryman, 1966; Vité
and Gara, 1962). These observations suggest that these
predators may utilize the chemical messenger(s) produced by
the bark beetle to find high prey densities. The ternary
mixture also attracted the bark beetle predator Temnochila

308

Table III

Numbers of Enoclerus lecontei Trapped in Response to
Synthetic Pheromones Presented Individually and in
Combination (Madera County, California, September 1967)

Treatment	Numbers trapped[1]			
	Sept. 20	Sept. 26	Sept. 21	Sept. 25
I, II, and III	10	2	4	2
I	3	3		
II	0	0		
III	3	5		
I and II			0	1
I and III			2	1
II and III			0	1
Infested bolt	2	6	0	5
Blank	0	0	0	0

[1] Numbers trapped each day represents the sum of 2
replications.

virescens chlorodia (Mannerheim) (Coleoptera: Ostomidae),
but only in small numbers (3 ♂♂, 1 ♀).

With one exception (see below), the chemotaxis exhibited
by pedestrian beetles exposed to male pheromone was an indicator of the flight response to the same olfactory stimuli.
A direct correlation between the walking and flight response
is crucial if a laboratory bioassay is used to monitor the
isolation and identification of bark beetle pheromones. The
field response to these attractants indicates the dominance
of olfactory stimuli during the aggregation of this bark
beetle. Also, these beetles are responding to extraordinar-
ily low concentrations of attractant, i.e., 1.5 mg of

compound I, 1 mg of II, and 0.5 mg of III (compounds I and
III are mixtures of both (+) and (-) isomers) delivered
over a 6-hr test period.

While field testing compounds isolated from the frass
of I. confusus, a strong response (34 ♀♀, 3 ♂♂) by I. latidens (LeConte) to the mixture of compounds I and II and a
weak response (5 ♀♀) to compound I was observed (Wood et al.,
1967a). These results were subsequently verified in a laboratory bioassay with the synthetic compounds. However,
when compound III was added to the attractive mixture of compounds I and II, the response from I. latidens was eliminated. I. latidens did not respond to I. confusus frass, extracts of frass, or the ternary synthetic mixture either
in the laboratory or the field. The mixture of I and II
evoked a strong response from I. confusus in the laboratory
bioassay but was totally inactive to I. confusus in the
field. The unusual synergistic system obtained by adding
compound III to the mixture of I and II, which blocks or
masks the I. latidens response and evokes a strong I. confusus response, poses some challenging chemical and biological problems. We are at present comparing the chemical
fractions of I. latidens frass with those of I. confusus.

IV. Specificity

Sex pheromones are an important isolating mechanism,
and a study of their specificity should contribute to a
more enlightened view of insect biosystematics and evolution. Now that the compounds produced by bark beetles are
susceptible to chemical identification (Silverstein et al.,
1966a), progress in this field should accelerate.

Pheromone specificity has been studied in closely related species of Ips in Hopping's (1963) species group IX
with largely allopatric distributions and in more distantly
related species from groups III, IX, and X with sympatric
distributions (Wood and Lanier, 1968). Populations of
these species were collected from infested host trees and
transported to the Berkeley laboratory where they were confined for rearing. Males were introduced into preformed
entrance tunnels on bolts cut from live trees of a natural
host species: I. confusus, I. plastographus (LeConte), and
I. calligraphus (Germar) on ponderosa pine; I. montanus
(Eichoff) on western white pine; and a new species (Lanier,
1967) on single leaf pinyon, P. monophylla Torr. and Frem.

CONTROL OF INSECT BEHAVIOR

The resultant attractant-laden frass was collected in gelatine capsules affixed over the entrance tunnel. Specificity was evaluated by exposing groups of female beetles to comparable quantities of frass from males and/or extracts of frass in the olfactometer. The male-produced materials were presented both separately and simultaneously. Where the attractants of 2 species were compared simultaneously, the positions in the olfactometer were reversed mid-way through the experiment. Species integrity of I. montanus, I. confusus, and the new species in group IX has been confirmed by interbreeding studies (Lanier, 1966, 1967).

Female I. plastographus, I. confusus, and I. calligraphus responded at a much lower level to frass produced by males of species in the 2 different species groups than to frass produced by males of their own species (Table IV). Female I. confusus responded at the same level to frass of male I. confusus, I. montanus, and the new species when each was presented separately (Table V). Females of I. montanus and the new species responded similarly. However, I. confusus

Table IV

Response of Ips Females to Male Pheromones Produced by Species in Various Species Groups

Species tested (♀♀)	Group No.	Source of frass (♂♂)	Group No.	Numbers exposed	Percent positive response
plastographus	III	plastographus	III	131	76.3
		confusus	IX	25	12.0
		calligraphus	X	47	17.0
confusus	IX	confusus	IX	776	59.7
		plastographus	III	143	9.8
		calligraphus	X	56	16.0
calligraphus	X	calligraphus	X	81	64.2
		plastographus	III	42	11.9
		confusus	IX	55	9.1

311

Table V

Response of Ips Females in Group IX to Male-produced
Pheromones Presented Separately

Species tested (♀♀)	Source of frass (♂♂)	Numbers exposed	Percent positive response
confusus	confusus	776	59.7
	montanus	487	51.1
	new species	494	57.9
montanus	montanus	275	42.6
	confusus	186	51.8
	new species	180	53.3
new species	new species	166	41.6
	confusus	80	41.3
	montanus	78	45.9

females showed a very highly significant preference for
male I. confusus frass when it was presented simultaneously
with frass from either I. montanus or the new species (Table
VI). Both I. montanus and the new species failed to show
such discrimination for their own frass. In the absence
of its own frass, I. confusus females did not discriminate
between the pheromones of other species. In the field
frass of I. confusus was about twice as attractive to I.
grandicollis (Eichhoff) (also in group IX) as were logs
infested by the latter species (Vité et al., 1964).

Ips confusus, I. montanus, and the new species were
exposed to the I. confusus synthetic pheromones in the
olfactometer (unpublished data). The individual compounds
evoked only an occasional trace response from the 3 species
at the following levels: 20 µg of compound I, 10 µg of II,
and 10 µg of III. As in previous studies, mixtures of
compound I + II, I + III, and I + II + III were highly
attractive to both sexes of I. confusus (Table VII). Only
the ternary mixture evoked a weak response from female
I. montanus. Females of the new species were moderately
responsive to mixtures of I + III and I + II + III. I. confusus responded to the ternary mixture at 1/100 to 1/10,000

CONTROL OF INSECT BEHAVIOR

Table VI

Response of Ips Females in Group IX to Male-produced
Pheromones of 2 Species Presented Simultaneously

Species tested (♀♀)	Source of frass (♂♂)	Numbers exposed	Percent positive response	Ratio
confusus	confusus montanus	689	51.7 18.0	2.9:1 ***
confusus	confusus new species	1010	41.4 22.5	1.8:1 ***
confusus	montanus new species	212	25.5 23.1	1.1:1
montanus	montanus confusus	234	32.9 23.5	1.4:1
montanus	montanus new species	173	17.9 15.0	1.2:1
new species	new species confusus	81	35.8 39.5	0.9:1

*** Difference is significant at the 0.001 % level.

of the level required to evoke a response from I. montanus.
The distributions of species within group IX are
largely allopatric but there are some areas of overlap
where populations are probably exposed to the pheromones of
another species in this group. For example, I. montanus is
found breeding primarily in western white pine which occurs
at high elevations in California. While I. confusus has
been recorded only occasionally from this host, as well as
from P. jeffreyi Grev. and Balf. and P. contorta Dougl., it
is not abundant at these elevations. Interspecific attraction would promote inviable matings, but preference of each
species for its own pheromones would reduce such interbreeding pressure. Because the secondary attractant can be produced in the non-host species occurring in the same forest
stand as the host (Jantz and Rudinsky, 1965; Wood et al.,

313

Table VII

Response of Male and Female *Ips* in Group IX to Synthetic *I. confusus* Pheromones

		Response levels[1]					
		confusus		new sp.		montanus	
Compounds	Amounts (μg)	♂♂	♀♀	♂♂	♀♀	♂♂	♀♀
I + II	20, 10	S	S	0	T	T	T
I + III	20, 10	M	S	T	M	T	0
II + III	10, 10	T	T	T	W	0	0
I + II + III	20, 10, 10	M	S	T	M	T	W
I + II + III	0.2, 0.1, 0.1	W	S			0	0
I + II + III	0.002, 0.001, 0.001	W	W				

[1] These response levels were established by the sequential release of several groups of 10 beetles and averaging the numbers of beetles exhibiting a positive response. 0=no response, T=trace (0-1 positive), W=weak (2-3), M=moderate (4-7), S=strong (8-10).

1966), initial recognition of the host species must be an effective isolation mechanism. Sympatric species that breed in the same individual host generally occur in different species groups. Breeding isolation is maintained in these species primarily by the specificity of their pheromones. An understanding of the biochemical mechanisms of pheromone specificity might be advanced by comparing the compounds produced by *I. confusus* with the compounds produced by an allopatric species (*e.g.*, *I. montanus*) and by a sympatric species (*e.g.*, *I. pini*).

V. References

Anderson, R. F. (1948). J. Econ. Ent. 41, 596.
Bakke, A. (1967). Z. Angew. Ent. 59, 49.
Berryman, A. A. (1966). Can. Ent. 98, 519.
Borden, J. H. (1967). Can. Ent. 99, 1164.
Borden, J. H., and Wood, D. L. (1966). Ann. Ent. Soc.
    Amer. 59, 253.
Cameron, E. A., and Borden, J. H. (1967). Can. Ent. 99,
    236.
Chapman, J. A. (1966). Can. Ent. 98, 50.
Chararas, C. (1966). C.R. Hebd. Séanc. Acad. Sci., Paris
    262, 2492.
Dyer, E. D. A., and Taylor, D. W. (1968). Can. Ent. 100,
    769.
Goeden, R. D., and Norris, D. M., Jr. (1964). Ann. Ent.
    Soc. Amer. 57, 141.
Hopping, G. R. (1963). Can. Ent. 95, 508.
Jantz, O. K., and Rudinsky, J. A. (1965). Can. Ent. 97,
    935.
Lanier, G. N. (1966). Can. Ent. 98, 175.
Lanier, G. N. (1967). Ph.D. thesis, Univ. of Calif.,
    Berkeley.
McCambridge, W. F. (1967). Ann. Ent. Soc. Amer. 60, 920.
McMullen, L., and Atkins, M. V. (1962). Can. Ent. 94, 1309.
Meyer, H. J., and Norris, D. M. (1967). Ann. Ent. Soc.
    Amer. 60, 642.
Person, H. L. (1931). J. Forestry 29, 696.
Pitman, G. B. (1966). Contrib. Boyce Thompson Inst. Pl.
    Res. 23, 147.
Pitman, G. B., and Vité, J. P. (1963). Contrib. Boyce
    Thompson Inst. Pl. Res. 22, 221.
Pitman, G. B., Kliefoth, R. A., and Vité, J. P. (1965).
    Contrib. Boyce Thompson Inst. Pl. Res. 23, 13.
Pitman, G. B., Renwick, J. A. A., and Vité, J. P. (1966).
    Contrib. Boyce Thompson Inst. Pl. Res. 23, 243.
Pitman, G. B., Vité, J. P., Kinzer, G. W., and Fentiman,
    A. F. (1968). Nature 218, 168.
Renwick, J. A. A. (1967). Contrib. Boyce Thompson Inst.
    Pl. Res. 23, 355.
Renwick, J. A. A., Pitman, G. B., and Vité, J. P. (1966).
    Naturwissenschaften 53, 83.
Rudinsky, J. A. (1963). Contrib. Boyce Thompson Inst. Pl.
    Res. 22, 23.

DAVID L. WOOD

Rudinsky, J. A., and Daterman, G. E. (1964). Can. Ent. 93, 1339.
Silverstein, R. M., Brownlee, R. G., Bellas, T. E., Wood, D. L., and Browne, L. E. (1968). Science 159, 889.
Silverstein, R. M., Rodin, J. O., and Wood, D. L. (1966a). Science 154, 509.
Silverstein, R. M., Rodin, J. O., Wood, D. L., and Browne, L. E. (1966b). Tetrahedron 22, 1929.
Tsao, C. H., and Yu, C. C. (1967). J. Georgia Ent. Soc. 2, 13.
Vité, J. P. (1965). Naturwissenschaften 52, 267.
Vité, J. P. (1967). Science 156, 105.
Vité, J. P., and Gara, R. I. (1962). Contrib. Boyce Thompson Inst. Pl. Res. 21, 251.
Vité, J. P., and Pitman, G. B. (1968). Nature 218, 169.
Vité, J. P., Gara, R. I., and Kliefoth, R. A. (1963). Contrib. Boyce Thompson Inst. Pl. Res. 22, 39.
Vité, J. P., Gara, R. I., and von Scheller, H. D. (1964). Contrib. Boyce Thompson Inst. Pl. Res. 22, 461.
Wilkinson, R. C. (1964). Florida Ent. 47, 57.
Wood, D. L. (1962). Pan-Pacif. Ent. 38, 141.
Wood, D. L., and Bushing, R. W. (1963). Can. Ent. 95, 1066.
Wood, D. L., and Lanier, G. N. (1968). Unpublished data.
Wood, D. L., and Vité, J. P. (1961). Contrib. Boyce Thompson Inst. Pl. Res. 21, 79.
Wood, D. L., Browne, L. E., Bedard, W. D., Tilden, P. E., Silverstein, R. M., and Rodin, J. O. (1968). Science 159, 1373.
Wood, D. L., Browne, L. E., Silverstein, R. M., and Rodin, J. O. (1966). J. Insect Physiol. 12, 523.
Wood, D. L., Stark, R. W., Silverstein, R. M., and Rodin, J. O. (1967a). Nature 215, 206.
Wood, D. L., Silverstein, R. M., and Rodin, J. O. (1967b). Science 156, 105.

ELECTROPHYSIOLOGICAL INVESTIGATION OF INSECT OLFACTION

Minoru Yamada

Faculty of Agriculture
Nagoya University
Nagoya, Japan

Table of Contents

I. Introduction

Progress in the study of isolation of biologically active substances is largely dependent on reliable bioassay methods. Classical methods utilizing the reflex response of animals produced by stimulation have been used successfully to study biologically active substances (Block, 1960; Butenandt and Hecker, 1961; Butenandt et al., 1959; Dethier, 1947, 1960; Dethier et al., 1952; Hamamura, 1959; Jacobson et al., 1960, 1961, 1963; Munakata et al., 1959). In addition, electrophysiological methods have been exploited, mainly by means of extracellular recordings from sensory organs (Boeckh, 1962, 1965; Boeckh et al., 1963, 1965; Lacher, 1964; Morita and Yamashita, 1961; Schneider, 1957a, b, 1962, 1963; Schneider et al., 1964, 1967; Yamada, 1966, 1967). These latter studies permitted us to supplement the results obtained by the behavioral methods. More importantly, it was possible to establish many functional details of unit cell responses which could not have been deduced solely from the reflex responses. The major part of the present paper is concerned with the progress of electrophysiology of insect olfaction. The results of recent studies conducted in the author's laboratory are also described briefly.

II. Basic Techniques

A. Stimulation

Evaluation of odor response of an organism, its olfactory organ, or of a unit cell is possible only when stimuli to be tested are qualitatively and quantitatively controlled. Many workers have tried to fulfill this requirement, but many points remain to be improved. In our qualitative experiments, the antenna was stimulated by squeezing a polyethylene bottle containing an odorous substance. The apparatus for stimulation in the quantitative experiments is shown in Fig. 1.

B. Recording

Summed receptor potentials of many cells elicited by odor puffs were easily recorded by inserting one electrode near the tip of the antenna and another electrode at the base. Such potentials were called the electroantennogram

(EAG) (Fig. 1). Extracellular recordings from single receptors could be made with a glass capillary electrode
inserted into a hair sensillum of the antenna and the indifferent electrode was placed in the hemolymph space of the
antenna. This method permitted simultaneous recordings of
receptor potentials and nerve impulses (Fig. 2).

For single unit recording from the antennal lobe, an
unanaesthetized cockroach was secured with pins and tape on
a cork plate, and a square section of the dorsal exoskeleton
of the frons was removed to expose the olfactory center. A
glass capillary electrode filled with 3M KCl was inserted
into the dorsal sensory neuron area of the deutocerebrum.

III. Peripheral Activity

A. Summed Receptor Potentials (EAG)

The EAG gives information on the qualitative spectrum
and the intensity-reaction relationship of a large number
of sensory units. It is a useful index for the bioassay
of biologically active substances (Boeckh et al., 1963;
Schneider, 1957a,b, 1962, 1963; Schneider et al., 1967).
For example, the sexual attractants extracted from any of
the female moths studied to date elicited strong EAG responses in the antennae of the male of the same species, but not
in the female (Boeckh et al., 1965; Schneider, 1962, 1963;
Schneider et al., 1967). However, in the case of the
American cockroach, Periplaneta americana L., the sex attractant of the female elicited the EAG in the antennae of the
male, female, and nymphs of this species (Boeckh et al.,
1963). The antennae of the fruit-piercing moths, Oraesia
excavata Butler and Adris tyrannus amurensis Staudinger,
gave very similar EAG responses to their food attractant as
well as to their repellent (Yamada, 1966) (Fig. 3).

B. Response from Single Receptor Cells

Dethier (1941) studied the sensilla of the antenna of
lepidopterous larvae, and assumed that these "large" sensilla basiconica were olfactory organs. Morita and Yamashita
(1961) recorded receptor potentials directly related to the
discharge of impulses from this "large" sensillum basiconi-
cum. Electrical responses to various chemicals have since
been reported in many insects by several workers. Many

MINORU YAMADA

receptor cells at rest fire spontaneously. These spontaneous action potentials, although recorded extracellularly
(Fig. 4), are diphasic (positive to negative) and resembled
those occurring after chemical stimulation.

1. Qualitative Spectrum of Receptor Cells

Olfactory stimuli increase the frequency of impulses,
depress spontaneous discharges, or cause no effect at all.
All of the single receptor cells studied so far are divided
into two main groups, called "specialists" and "generalists".
    Typical odor specialists are the sexual-attractant
receptors of male moths and honey bee drones. In the male
silkworm, Bombyx mori L., these cells respond strongly to
the female's sexual attractant, Bombykol, and to its geometrical isomers. Linalool and terpineol are much weaker
stimulants. This picture is reversed in the B. mori female:
the sexual attractant induces a slight reaction in a few
cases, while linalool and terpineol produce strong responses. It appears that the B. mori female lacks receptors to
sense its own lure substance (Table I). In addition to the
sexual odor specialists, food odor specialists were also
found in the carrion beetle (Boeckh, 1962), the blowfly and
the grasshopper (Boeckh et al., 1965).

Table I

Reaction Spectrum of the Hair Sensillum of
the Male and Female of Bombyx mori

Substance	Response[a] Male	Female
Bombykol (TC)	+++	(+)
Bombykol isomers	++	(+)
Cycloheptanon	+	(+)
Terpineol	+	++
Linalool	(+)	++

[a] Response rating: +, weak; ++, medium; +++, strong;
(+) rare. (From Priesner, unpublished data.)

320

Odor generalists are cells which respond to a variety of different chemicals. The sensilla of A. tyrannus amurensis responded with an increase in impulse frequency, not only to their food attractant but also to their repellent (Fig. 5).

Experiments with a dual stimulation, with ether vapor first and chloroform vapor second, or vice versa (Fig. 6), show that the second stimulus may control the electrical activity of the receptors in the sensillum by changing the nerve activity produced by the first stimulus (Yamada, 1967).

2. Quantitative Study of Receptors

A typical reaction of single Bombykol receptor cells in the male Bombyx antenna is graphically presented in Fig. 7. The curves of the EAG and of the generator potential almost fit each other. This supports the assumption that the EAG is the sum of many single receptor (generator) potentials. The single Adris receptor unit increases its impulse frequency as the strength of the odor stimulus increases (Fig. 8).

Reaction times (latencies) decrease with increasing stimulus intensity. Most of the olfactory receptor cells we have investigated so far belong to the phasic-tonic reaction type. They respond to a stimulus with a transient phase of very high frequency followed by a steady-state low frequency.

3. Explanation of Electrical Activity in the Olfactory Receptor

Insect olfactory receptors contain sensory neurons which appear to be similar in many respects to those of the labellar chemoreceptors of the blowfly (Morita, 1959; Morita and Yamashita, 1959a,b; Wolbarsht and Hanson, 1965) and the Pacinian corpuscle (Gray and Sato, 1953; Hunt and Takeuchi, 1962). They are bipolar cells, and the receptor membrane is modified to respond to a specific stimulus; spikes are diphasic in shape, positive to negative (Fig. 4); and there is a close time correlation between slow potentials and impulses, i.e., the increase in impulse frequency is always accompanied by a slow negative potential at the different electrode, while the decrease in impulse frequency is accompanied by a slow positive potential (Boeckh, 1962, 1965;

Boeckh et al., 1965; Morita and Yamashita, 1961; Yamada, 1966).

These results may be explained, as described for the labellar chemoreceptors of the fly and the Pacinian corpuscle, by the scheme illustrated in Fig. 9. For simplicity, only one neuron is shown in the hair. When odor molecules such as those of a sex attractant arrive at the tip of the olfactory hair, the membrane of the chemoreceptive surface (the shaded area) is depolarized. Current then flows along the course indicated by arrows. The potential drop due to this current is recorded as a sustained negativity by the electrode kept in contact with the space outside the neuron (receptor potential or generator potential). This current in turn initiates impulses in the neuron at the site near the hair base (area b). The additional current due to this sink flows as indicated by the thin arrows in Fig. 9B, and the potential drop due to this current is recorded by the same electrode as an increase in positivity superimposed upon the sustained negativity. The impulses are conducted proximally toward the central nervous system, and also distally toward the hair tip (indicated by the thick arrows in Fig. 9B). Thus, the distally-conducted impulse makes the recorded impulse briefer than the initiated impulse, and this generally results in a more or less diphasic impulse.

IV. Spike Activity in the Olfactory Bulb

There have been a few reports concerning the olfactory responses from single units in the central nervous system in invertebrates. Figure 10 shows representative examples of single unit responses from the male cockroach. Propionic acid induced a positive slow potential which was sustained during the stimulation, while the discharge of impulses was depressed. After the cessation of stimulation, the frequency of impulses increased (Fig. 10A). On the other hand, crude extracts of sex attractant (Fig. 10B) and methyl-ethyl ketone (Fig. 10C) produced a negative slow potential which is accompanied by a burst of activity. This was followed by a null period and then a slow increase to the background level. These results may be explained in terms of hyperpolarization or depolarization of the synaptic membrane. Both males and females were found to possess units sensitive to the sex attractant and other odors.

Figure 11 shows responses of the highly specific neurons

to odor stimuli. The responses, recorded from the same male as in Fig. 10, showed no spontaneous discharge (Figs. 11A, B). It responded vigorously to the sex attractant with a slow negativity which is accompanied by a train of impulses during the stimulation (Fig. 11A). However, it failed to respond to the following odor stimuli: propionic acid, acid benzyl acetate, p-dichlorobenzene, 2-methyl-2-butanol, acetophenone, geraniol, 1,2-dichloroethane, trichloroethylene, cycloheptanone, diethylsulphate, cyclopentanone, ethyl ether, methyl alcohol, and methyl-ethyl ketone. The responses recorded from a female (Figs. 11C,D) also showed a high specificity to the crude extract of the sex attractant (Fig. 11C). In males, one would expect strong responses of neurons in the deutocerebrum to crude extracts of the female sex attractants. The fact that the female possesses similar neurons that respond similarly to the crude extracts of sex attractants is somewhat astonishing.

Acknowledgments

The author wishes to express his deepest appreciation to Professors T. Tamura, K. Iyatomi, K. Munakata, T. Saito, and T. Narahashi, Duke University, for their valuable guidance and advice. He is also indebted to Professors M. Kuwabara and H. Morita, Kyushu University, for their valuable suggestions. The crude sex attractant from the American cockroach was kindly provided by Professor S. Ishii, Kyoto University.

V. References

Block, B. C. (1960). J. Econ. Entomol. 53, 172.
Boeckh, J. (1962). Z. Vergl. Physiol. 46, 214.
Boeckh, J. (1965). Proc. Second Internat. Symp. on Olfaction
    and Taste (Ed., T. Hayashi) Pergamon Press, Oxford. 721.
Boeckh, J., Kaissling, K. E., and Schneider, D. (1965).
    Cold Spring Harbor Symp. Quant. Biol. 30, 263.
Boeckh, J., Priesner, E., Schneider, D., and Jacobson, M.
    (1963). Science 141, 716.
Butenandt, A., Beckmann, R., Stamm, D., and Hecker, E.
    (1959). Z. Naturforsch. 14b, 283.
Butenandt, A., and Hecker, E. (1961). Angew. Chemie. 73,
    349.
Dethier, V. G. (1941). Biol. Bull 80, 403.
Dethier, V. G. (1947). "Chemical Insect Attractants".
    Blakiston, Philadelphia.
Dethier, V. G. (1960). J. Econ. Entomol. 53, 134.
Dethier, V. G., Hackly, B. E., and Wagner-Jauregg, T. (1952).
    Science 115, 141.
Gray, J. A. B., and Sato, M. (1953). J. Physiol. 122, 610.
Hamamura, Y. (1959). Nature 183, 1746.
Hunt, C. C., and Takeuchi, A. (1962). J. Physiol. 160, 1.
Jacobson, M., Beroza, M., and Jones, W. A. (1960). Science
    132, 1011.
Jacobson, M., Beroza, M., and Jones, W. A. (1961). J. Amer.
    Chem. Soc. 83, 4819.
Jacobson, M., Beroza, M., and Yamamoto, R. T. (1963).
    Science 139, 48.
Lacher, V. (1964). Z. Vergl. Physiol. 48, 587.
Morita, H. (1959). J. Cell Comp. Physiol. 54, 189.
Morita, H., and Yamashita, S. (1959a). Science 130, 922.
Morita, H., and Yamashita, S. (1959b). Mem. Fac. Sci.
    Kyushu Univ., Ser. E. (Biol.) 3, 81.
Morita, H., and Yamashita, S. (1961). J. Exp. Biol. 38, 851.
Munakata, K., Saito, T., Ogawa, S., and Ishii, S. (1959).
    Bull. Agric. Chem. Soc. Japan 23, 64.
Schneider, D. (1957a). Experientia 13, 89.
Schneider, D. (1957b). Z. Vergl. Physiol. 40, 8.
Schneider, D. (1962). J. Insect Physiol. 8, 15.
Schneider, D. (1963). Proc. First Internat. Symp. on
    Olfaction and Taste (Ed., Y. Zotterman) Pergamon Press,
    Oxford. 85.
Schneider, D., Block, B. C., Boeckh, J., and Priesner, E.
    (1967). Z. Vergl. Physiol. 54, 192.

Schneider, D., Lacher, V., and Kaissling, K. E. (1964). Z. Vergl. Physiol. 48, 632.

Wolbarsht, M. L., and Hanson, F. E. (1965). J. Gen. Physiol. 48, 673.

Yamada, M. (1966). Insect Toxicologists' Information Service 9, 191.

Yamada, M. (1967). Appl. Ent. Zool. 2, 22.

Yamada, M. (1968a). Nature 217, 778.

Yamada, M. (1968b). Botyu-Kagaku 33, 37.

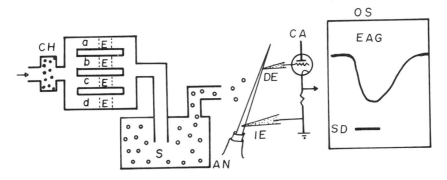

Figure 1.--Diagrammatic illustration of the experimental arrangement. CH: charcoal; a, b, c, d: glass tubing of various dimensions; E: electromagnetic valve; S: test bottles containing a stimulus substance; AN: antenna; DE: different electrode; IE: indifferent electrode; CA: cathode follower; OS: oscilloscope; SD: stimulus duration; EAG: electroantennogram.

Stimulus intensity is controlled by means of glass tubing of various dimensions through which air is passed.

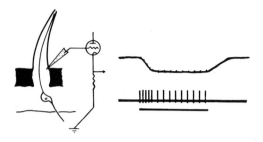

Figure 2.--Method of recording receptor potentials and
    action potentials from a single sensillum basiconicum.
    Upper (D-C) trace shows receptor potential; lower (R-C)
    trace shows nerve impulses.

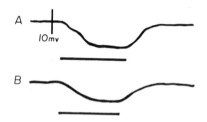

Figure 3.--EAG-responses of Adris tyrannus amurensis to
    A: food attractant (odor of grape); B: repellent
    (LNY-13) (from Yamada, unpublished data).

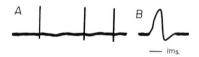

Figure 4.--DC recording of spikes from olfactory receptor
    cells. A: slow sweep record; B: fast sweep record.

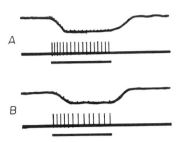

Figure 5.--<u>Adris tyrannus amurensis</u>, male. A: response to
food attractant (DC recording). B: response to repellent (AC recording). Black bar indicates the stimulation.

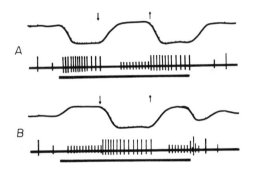

Figure 6.--Antennal responses to dual stimulation with ether
and chloroform. A: horizontal black bar represents
the stimulation by ether. Arrows, ↓ and ↑, indicate
the start and end of chloroform stimulation which was
superimposed on the ether stimulation. B: horizontal
black bar indicates the chloroform stimulation.
Arrows, ↓ and ↑, indicate the start and end of ether
stimulation, which was superimposed on the chloroform
stimulation. (From Yamada, 1967).

327

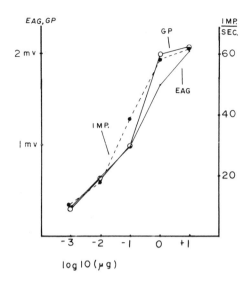

Figure 7.--<u>Bombyx mori</u>, male. Comparison of the Bombykol
    EAG-reaction with the generator potential (GP) and the
    mean impulse frequency (IMP.) during the first 500
    msec of the response. Abscissa: Bombykol concentration.
    The values of GP and IMP. are averages of 3-6 measure-
    ments (From Priesner, unpublished data).

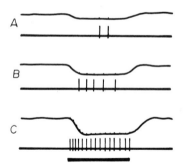

Figure 8.--<u>Adris tyrannus amurensis</u>, male. Responses of a
    long hair sensillum to a series of increasing concen-
    trations of the odor of grape. Upper trace: DC record-
    ing. Lower trace: AC recording. (From Yamada, unpub-
    lished data.)

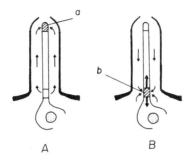

Figure 9.--Schematic drawing of an olfactory hair illustrat-
ing olfactory-reception. Shaded areas represent the
active regions of the olfactory neuron. A: generator
(receptor) potential; B: impulse initiation; a:
olfactory receptor region; b: site of initiation of
impulses. Thick arrows represent the direction along
which impulses are conducted; thin arrows represent
current flow. (Courtesy of Dr. H. Morita.)

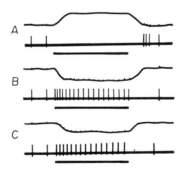

Figure 10.--Response of a single unit in the antennal lobe
of male Periplaneta americana. A: to propionic acid;
B: to crude extract of sex attractant; C: to methyl-
ethyl ketone. (From Yamada, 1968a.)

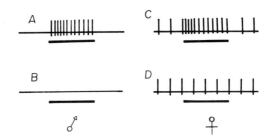

Figure 11.--Responses of specific neurons in the antennal
lobe of male and female <u>Periplaneta</u> <u>americana</u>. A,C:
to the crude extract of the female sex attractant; B,D:
to methyl-ethyl ketone. (From Yamada, 1968b.)

HOST ATTRACTANTS FOR THE RICE WEEVIL AND THE CHEESE MITE

Izuru Yamamoto and Ryo Yamamoto

Department of Agricultural Chemistry
Tokyo University of Agriculture
Setagaya, Tokyo, Japan

Table of Contents

I. Introduction

Methods of insect control which produce minimum contamination of foods and feeds, water, soil, wildlife, and the
environment as a whole have been given careful reconsidera-
tion in recent years. Insights into the behavioral responses
of insects to chemicals, from both their food and the
insects themselves, have indicated their potential for control in the field of agricultural entomology. However, very
few studies have been carried out with stored product insects, especially concerning the chemical nature of the
chemosensory stimulants. There is evidence that some stored
product insects respond to stimuli from components in their
natural food and from fungi (Loschiavo, 1965a).

Since 1965, our group in the Tokyo University of Agriculture has been investigating the constituents of rice and
cheese that attract the rice weevil and cheese mite, respectively. In Japan, many important food products, such as
cheese, spices, bean-paste ('Miso'), dried fish, powdered
milk, chocolate, and powdered grain, are infested by mites,
causing serious losses and sanitary problems. Tyrophagus
putrescentiae (Schrank) is an important mite and we have
chosen to study cheddar cheese as a source of attractants.
Preliminary experiments showed that this mite was attracted
to cheese by olfaction. The rice weevil, Sitophilus zeamais
Motschulsky, is an important stored product insect, widely
infesting rice and other grains and the products therefrom.
There was an indication that an olfactory stimulant might
be present in rice. The studies reported here are aimed
at the eventual isolation of these attractants.

II. Rice Weevil Attractants from Rice Grain

A. Bioassay

We devised a modified Chamberlain (1956) olfactometer
for the laboratory bioassay (Fig. 1). This olfactometer
seemed to permit detection of the attractant according to
the definition of Dethier et al. (1960). The sample was
placed or absorbed on one of the filter papers, 9 cm dia and
with many pin-holes. The bioassay was conducted in darkness,
at 25-30°C and 14-16% relative humidity, with rice weevils
that had been starved for 2 days prior to a test. Fifty
unsexed weevils, which had emerged 4 days earlier from rice
grains, were placed in a cage and provided with a choice

between a sample side and a control side. The numbers of insects at the sample side were counted after 10 min. Air flow to each side was reversed, and counts were repeated at 10 min intervals until 6 counts were obtained. The mean numbers of insects at the sample side before and after switching of the flow were recorded. When the mean number was >39, attractivity was evaluated as +3, 35-38 as +2, 31-34 as +1, and 20-30 as 0. The total value (obtained by adding the five evaluations) was used for scoring the attractivity.

Continuous observation of insect aggregation was also attempted by viewing through a hole with minimal light. These observations were made for a sample of the acidic fraction: during the first few minutes, the insects located around the center of the cage, and after 6 min they started to migrate to the sample side. After 8 min, only 9 of 50 insects remained at the control side. The flow of the sample air was switched to the other side after 10 min and the behavior observed during the first 10 min was repeated. The migration, however, became slow and indistinct after 30 min. The migration pattern was different for the different types of attractant samples. For the ether extract, migration to the sample side occurred almost regularly upon switching of the flow, whereas the migration was noticeable only during the first 10 min for the non-acidic fraction.

The bioassay was performed under low humidity conditions using starved rice weevils. However, the attractive fraction and the non-attractive fraction demonstrated by the olfactometer also gave the same results using a modified Munakata odor trap method (Munakata et al., 1959) and a modified Loschiavo petri dish method (Loschiavo, 1965b), which partly simulated natural conditions.

B. Fractionation

The procedure adopted for concentrating the main attractive component of rice grain is illustrated on the following page. This was arrived at after attempting several methods. We were able to concentrate the attractive fraction to 0.0024% of the starting material.

C. Properties of the Attractive Component

Since the purity of the acidic fraction was inadequate

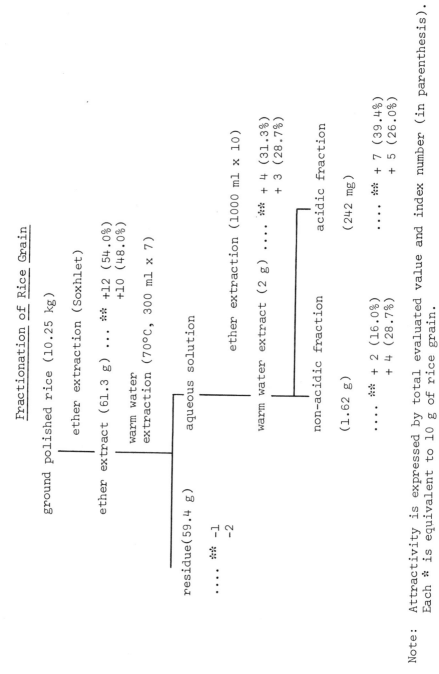

Fractionation of Rice Grain

ground polished rice (10.25 kg)

    ether extraction (Soxhlet)

ether extract (61.3 g) .... ※ +12 (54.0%)
                      +10 (48.0%)

    warm water
    extraction (70°C, 300 ml x 7)

residue(59.4 g)   aqueous solution

..... ※ -1
     -2
                  ether extraction (1000 ml x 10)

         warm water extract (2 g) ..... ※ + 4 (31.3%)
                                 + 3 (28.7%)

non-acidic fraction   acidic fraction

(1.62 g)   (242 mg)

..... ※ + 2 (16.0%)   ..... ※ + 7 (39.4%)
      + 4 (28.7%)   + 5 (26.0%)

Note: Attractivity is expressed by total evaluated value and index number (in parenthesis).
       Each ※ is equivalent to 10 g of rice grain.

to allow meaningful interpretations of the spectra or other physical constants, we attempted to learn the type of molecule responsible for attractivity from the partition behavior and the changes in attractivity after reaction with various chemical reagents. The main attractive component seemed to be stable to heat (150°C, 1 hr), ultraviolet light (2537 or 3650 Å, 2 hr) and base (ethanolamine or 1 N aqueous KOH for 2 weeks at room temperature; 3% alcoholic KOH, refluxing for 1 hour), non-volatile at 150°C, and acidic. The reasoning for assigning the carboxylic nature to the component was as follows: the attractivity was taken into bicarbonate solution from the ether solution and back-extracted to the ether phase after acidification; methylation with diazomethane gave complete loss of attractivity, but the attractivity was recovered after saponification (Table I). To show the property of partitioning into water, the attractant molecule should be provided with suitable hydrophilic functions besides a carboxyl group. Since the attractivity can also be taken into ether, the nature and the number of the hydrophilic functions are naturally limited. A carbonyl group is a possibility, because of the disappearance of attractivity by reaction with sodium borohydride. The non-effect of 3,5-dinitrobenzoyl chloride probably indicates the absence of acylable function such as hydroxyl. Further work should be carried out to isolate a pure compound, but this information is useful in suggesting isolation procedures.

D. Synergism

We have attempted to isolate the attractant with various types of chromatography. In each case, the attractivity was either not recovered or poorly recovered from the fraction separated. The instability of the material may not be the cause. A possibility is synergism between two or more fractions to exert attractivity. Several cases have been reported in the literature in which reduction of attractivity was observed as the isolation proceeded.

The disappearance of attractivity from the individual fractions when the ether extract was further fractionated with formamide, and the recovery of the attractivity on recombination of the separated fractions (synergism), is illustrated on the next page.

This information illustrates the complexity of the attractivity phenomenon. None of 38 common pure organic

An Example of Synergism

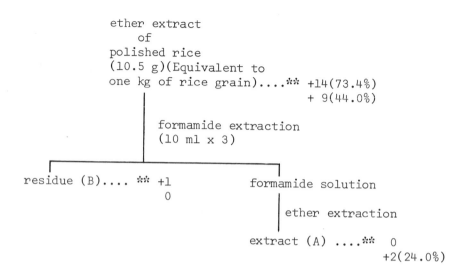

ether extract
    of
polished rice
(10.5 g)(Equivalent to
one kg of rice grain)....** +14(73.4%)
                           + 9(44.0%)

formamide extraction
(10 ml x 3)

residue (B).... ** +1
                    0

formamide solution

ether extraction

extract (A) ....** 0
                 +2(24.0%)

combination of .... ** +14(61.4%)
A and B   + 5(28.0%)

Note: Each * is equivalent to 10 g of rice grain.

acids and related compounds were comparable in attractivity
to the acidic fraction from rice grain. The reaction of
various chemical reagents with the crude attractive fraction
may also occur on a synergistic component included in the
fraction, thus decreasing the attractivity. This may lead
to misinterpretation of the nature of the attractant.

E. Other Attractants

Although our discussion mainly dealt with the ether
extract of the polished rice and the acidic fraction obtained from the warm water extract of this ether extract,
we shall comment on the other attractive components to
illustrate the complexity of the problem.
An attractive component from nitrogen gas aeration of
the rice grain was trapped by a dry-ice/acetone trap, but

Table I

Effect of Chemical Reagents on Attractivity to _Sitophilus zeamais_

Sample	Reagent	Attractivity: Before treatment	Attractivity: After treatment
Ether extract	CH_2N_2	** +12(54.0%) +14(56.0%)	** +1[a] +1
Warm water extract	CH_2N_2	** + 7(39.4%) + 5(26.0%)	** + 3(17.4%) ** 0
Acidic fraction	$NaBH_4$	**** + 8(39.4%)	**** +1(11.3%) -3(-18.7%)
Ether extract	3,5-dinitro-benzoyl chloride	** +11(50.6%) +12(58.0%)	** +10(48.0%) + 8(44.7%)

Each * equivalent to 10 g of rice grain.

a After hydrolysis ** +12(54.0%)
+ 5(31.3%)

not by a bicarbonate solution trap; thus we have a non-acidic substance. Part of the attractivity was extracted directly from grain with n-hexane. When the ether extract was concentrated, weak attractivity was detected in the distillate, but most of the activity remained in the residue. On the other hand, when the water extract of the ether extract of rice grain was concentrated under reduced pressure, a part of the attractivity appeared in the distillate. This steam-distillable attractant was acidic, and thus differs from the aerated attractant. There seemed to be volatile and non-volatile acidic attractants.

As a result of this work, the chemical basis of the attractivity of rice weevil to the host rice grain was demonstrated. Further work toward eventual isolation of the attractive materials is in progress.

III. Cheese Mite Attractants from Cheese

A. Bioassay

Using a modified McIndoo's Y-tube olfactometer, we examined various factors in the bioassay. The presence of cheese and the starvation of the mites were the most important factors for causing migration of the mites. This suggested the presence of an attractive component in the neutral fraction of the steam distillate of cheddar cheese. However, this method was not satisfactory for monitoring various fractionation procedures. A new odor trap method was devised, which greatly facilitated the progress of the work. Two folded filter papers (5 x 30 mm) loaded with an amount of test material were placed in 2 tubes. Another 2 tubes contained filter papers without any test material or with material to be compared. The 4 tubes were arranged in a petri dish (15 x 4 cm) at right angles to one another with the open end of the tubes facing away from the center of the dish. Approximately 2,000 cheese mites were introduced in the center of the petri dish, which was then placed on a basin filled with water and covered with a dark cloth for 1 hour at room temperature. The mites which migrated to each tube were killed by adding hot water and counted under a microscope. The degree of attractivity is expressed as follows:

$$\text{Attractivity \%} = \frac{\text{Treatment}}{\text{Treatment + Control}} \times 100$$

When test material was also placed in the control tubes, relative potency could be shown (crosstest). All stages and both sexes of mites, which had been starved for 1 day prior to the test, were used. When introduced, the mites distributed themselves uniformly on the surface of the dish within 20 - 30 minutes and migrated into the tubes within 40 - 50 minutes. Statistically, a response >65% is significantly associated with the presence of attractivity, but we adopted 70% or more as an indication of attractivity.

B. Fractionation

Cheese was blended with water and lyophilized under high vacuum or steam distilled under reduced pressure (50-60°C/80-100 mm Hg). The distillate was attractive and further fractionated into the basic, acidic and neutral fractions as shown on the following page. The main cheese mite attractant exists in the neutral fraction of the volatile material.

C. Properties of the Attractive Component

Although the neutral fraction was a mixture, showing about 50 peaks on a gas chromatogram, the effect of various chemical reagents on the attractivity of the crude material was investigated to determine the general nature of the attractant. When the CD-IV fraction was fractionated with Girard T reagent, the main attractivity appeared in the non-carbonyl fraction and the minor one in the carbonyl fraction. Tollens' reagent and sodium borohydride seemed to reduce attractivity slightly. In a sealed tube with aqueous 10% NaOH at 95°C for 2 hr, CD-IV and the non-carbonyl fraction did not seem to lose attractivity, indicating the non-ester nature of the attractant. The attractant was stable in neutral to alkaline permanganate for at least 3 hr at room temperature. The effect of diazomethane on the non-carbonyl fraction was inconclusive, probably partly destroying the attractivity. Remarkable reduction of the attractivity of both CD-IV and the non-carbonyl fraction was obtained when reacted with acylating reagents, including acetyl chloride, 3,5-dinitrobenzoyl chloride, phenylisocyanate and α-naphthyl isocyanate. The crude product from the reaction with the dinitrobenzoyl chloride regained attractivity by saponification with 10% methanolic KOH at 95°C.

Fractionation of Cheese

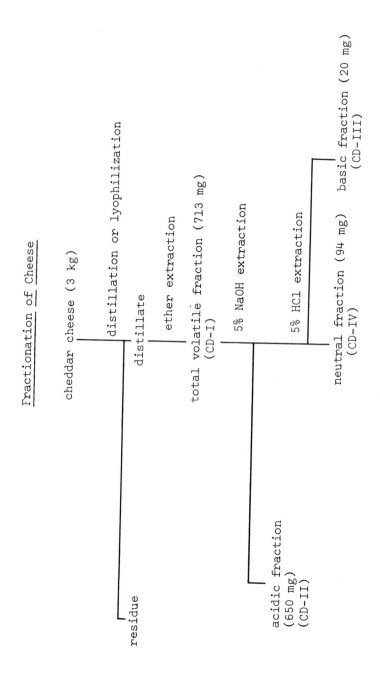

cheddar cheese (3 kg)

distillation or lyophilization

distillate

ether extraction

total volatile fraction (713 mg)
(CD-I)

5% NaOH extraction

5% HCl extraction

residue

acidic fraction
(650 mg)
(CD-II)

neutral fraction (94 mg)
(CD-IV)

basic fraction (20 mg)
(CD-III)

Thus, the main attractive component seemed to be a neutral non-carbonyl compound having acylable group.

D. Isolation

1. Attempted separation of the attractant as 3,5-dinitrobenzoate. The CD-IV (neutral) fraction gave a mixture of 3,5-dinitrobenzoates by reaction with benzoyl chloride. Thin-layer chromatographic separation gave a benzoate of mp 90°C, which showed attractivity when hydrolyzed. The benzoate was characterized as the ethanol derivative. However, ethanol itself or the hydrolyzate of the ethanol benzoate did not show attractivity. The attractant benzoate seemed present as a minute concomitant of the ethanol benzoate.

2. Nitrogen gas aeration. The CD-IV fraction was aerated with nitrogen at room temperature for several hours, then at 50-70°C for 1 day. The whole volatile fraction (about 1/3 of CD-IV) was attractive. The nonvolatile part obtained by this procedure showed only slight attractivity. This procedure efficiently removed the contaminant of higher retention time on gas chromatography and avoided contamination of the column.

3. Column chromatography. The aerated fraction from CD-IV was subjected to silica gel column chromatography using n-pentane with increasing ratios of ether as the eluting solvent. The eluate with 20% ether in n-pentane gave the highest attractivity.

4. Gas chromatography. The CD-IV fraction was gas chromatographed (Fig. 3A) and the eluates were cut into 4 fractions and bioassayed. The second fraction, with a retention time of 20-40 min, gave major attractivity (Table II), the chromatogram of this region being complex in pattern. However, when the CD-IV fraction was aerated and column chromatographed in advance as described above, the gas chromatogram of the attractive fraction showed a much simpler pattern of peaks (Fig. 3B). The amount of the attractive fraction thus obtained was estimated at less than 10^{-4}% of the cheese. Further separation is under way, starting from about 100 kg of cheese.

Table II

Attractivity of Gas Chromatographic
Fractions to T. putrescentiae

| Fraction | Attractivity | | |
(retention time in min)	5 µg	50 µg	100 µg
0 - 20	–	–	+
20 - 40	+	+	+
40 - 60	–	–	+
60 - 100	–	–	+

+ Responded
- No response

Acknowledgments

This research has been supported by a grant from the Unites States Department of Agriculture, ARS, under PL 480. Co-workers are Dr. Hiroshi Honda, Messrs. Takumi Yoshizawa, Hideaki Sasada, Tadashi Nittono, and Fumio Mori.

IV. References

Chamberlain, W. F. (1956). J. Econ. Entomol. 49, 659.
Dethier, V. G., Browne, L. B., and Smith, C. N. (1960). J. Econ. Entomol. 53, 134.
Loschiavo, S. R. (1965a). Proc. Entomol. Soc. Manitoba 21, 10.
Loschiavo, S. R. (1965b). Ann. Entomol. Soc. Amer. 58, 383.
Munakata, K., Saito, T., Ogawa, S., and Ishii, S. (1959). Bull. Agr. Chem. Soc. Japan 23, 64.

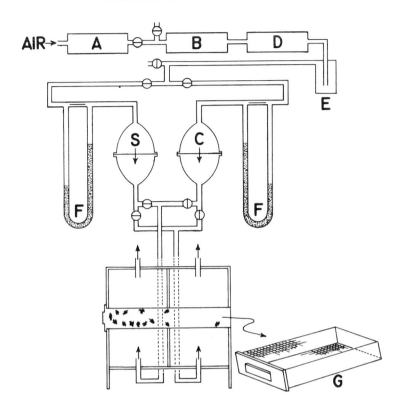

Figure 1.--Olfactometer used to monitor the fractionation of attractants for S. zeamais in rice grain.
A: compressor; B: integral flow-meter; D: charcoal purifier; E: humidity regulating vessel; F: flow-meter; S: filter paper holder (with sample), C: (without sample); G: insect cage (16 x 8 x 3 cm). → indicates flow of air.

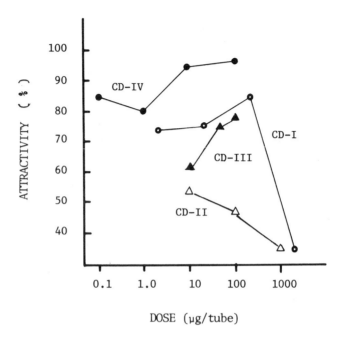

DOSE (μg/tube)

Figure 2.--Dose - attractivity of cheese fractions to T. putrescentiae. Dose equivalent to 300 mg of cheese: CD-I, 70 μg; CD-II, 65 μg; CD-III, 2 μg; CD-IV, 9 μg.

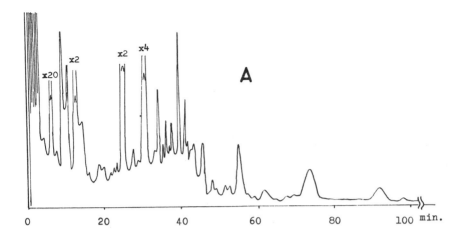

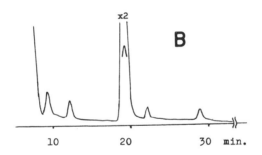

Figure 3.--Gas chromatograms of volatile, neutral fraction
   from cheese. A: CD-IV fraction. Conditions: 2 m x
   4 mm ID; column packed with 20% PEG-6000 on 60 - 80
   mesh chromosorb W; column temperature 70°C for 20 min,
   programed at 5°C/min to 170°C; flow rate 23 ml N_2/min.
   B: CD-IV after nitrogen aeration and column chromatography. Conditions: 2 m x 4 mm ID; packed with 20%
   PEG-20M on 60 - 80 mesh chromosorb W; 90°C for 10 min,
   programed at 2.5°C/min to 200°C; 23 ml N_2/min.